Taxonomic Revision of the Ant Genus *Linepithema* (Hymenoptera: Formicidae)

Alexander L. Wild

Taxonomic Revision of the Ant Genus *Linepithema* (Hymenoptera: Formicidae)

Alexander L. Wild

University of California Press

Berkeley • Los Angeles • London

University of California Press, one of the most distinguished university presses in the United States, enriches lives around the world by advancing scholarship in the humanities, social sciences, and natural sciences. Its activities are supported by the UC Press Foundation and by philanthropic contributions from individuals and institutions. For more information, visit www.ucpress.edu.

University of California Publications in Entomology, Volume 126
Editorial Board: Bradford A. Hawkins, John Heraty, Lynn S. Kimsey, Serguei V. Triapitsyn, Penny Gullan, Philip S. Ward, Kipling Willy

University of California Press
Berkeley and Los Angeles, California

University of California Press, Ltd.
London, England

Library of Congress Cataloging-in-Publication Data

Wild, Alexander L., 1973-
  Taxonomic revision of the ant genus Linepithema (Hymenoptera:Formicidae) / Alexander L. Wild.
     p. cm. -- (University of California publications in entomology ; v. 126)
  Includes bibliographical references.
  ISBN 978-0-520-09858-9 (pbk. : alk. paper)
  1. Linepithema--Classification. I. Title.
  QL568.F7W628 2007
  595.79'6--dc22
                          2007000110

The paper used in this publication meets the minimum requirements of
ANSI/NISO Z39.48-1992 (R 1997) (Permanence of Paper).{∞}

CONTENTS

ACKNOWLEDGEMENTS

Phil Ward, Jack Longino, Roy Snelling, Penny Gullan, Beto Brandão, Barry Bolton, and Jo-anne Holley made valuable comments on the manuscript. I would like to thank Bolivar R. Garcete-Barrett, the U. S. Peace Corps, and the Fundación Moises Bertoni for their help in Paraguay. Juan Vieira and David Donoso (QCAZ) provided valuable assistance in Ecuador, as did Jack Schuster in Guatemala and Kelvin Guerrero in the Dominican Republic. Nicole Heller assisted in the field in Argentina. Fabiana Cuezzo (IFML), Roberto Brandão (MZSP), Giovanni Onore (QCAZ), Stefan Schoedl (NHMW), Roberto Poggi (MCSN), Bernhard Merz (MHNG), Daniel Burkhardt (NHMB), and Barry Bolton (BMNH) made possible visits to their respective museums. Stefan Cover, Jack Longino, Bill MacKay, Jes Pedersen, Leo Rivera, Thibaut Delsinne, Andy Suarez, Ted Schultz, Jarbas Queiroz, Jo-anne Holley and Roy Snelling provided me with useful material. Jo-anne Holley helped in the preparation of figures. This work was made possible by financial support from the United States National Science Foundation grant #0234691, by a National Science Foundation Graduate Research Fellowship, and by grants from the University of California at Davis: Jastro Shields, Center for Population Biology, and Center for Biosystematics.

ABSTRACT

The primarily Neotropical dolichoderine ant genus *Linepithema* is revised at the species level for the first time. Morphological and biogeographic data support the recognition of 19 species. The following taxonomic scheme is proposed: *L. angulatum* (Emery) **stat. nov.** [= *pordescens* (Wheeler) **syn. nov.**], *L. dispertitum* (Forel), *L. flavescens* (Wheeler & Mann) **stat. nov.**, *L. fuscum* Mayr, *L. gallardoi* (Brèthes) [= *breviscapa* (Santschi) **syn. nov.** = *impotens* (Santschi) **syn. nov.**], *L. humile* (Mayr) [= *arrogans* (Chopard) = *riograndense* (Borgmeier)], *L. iniquum* (Mayr) [= *bicolor* (Forel) **syn. nov.** = *dominicense* (Wheeler) **syn. nov.** = *fuscescens* (Wheeler) **syn. nov.** = *melleum* (Wheeler) **syn. nov.** = *nigellum* (Emery) **syn. nov.** = *succineum* (Forel) **syn. nov.**], *L. keiteli* (Forel) [= *subfaciatum* (Wheeler & Mann) **syn. nov.**], *L. leucomelas* (Emery) [= *aspidocoptum* (Kempf) **syn. nov.**], *L. micans* (Forel) **stat. nov.** [= *platense* (Forel) **syn. nov.** = *scotti* (Santschi) **syn. nov.**], *L. oblongum* (Santschi), *L. piliferum* (Mayr). Seven species are described as new: *L. anathema* **sp. nov.**, *L. aztecoides* **sp. nov.**, *L. cerradense* **sp. nov.**, *L. cryptobioticum* **sp. nov.**, *L. neotropicum* **sp. nov.**, *L. pulex* **sp. nov.**, and *L. tsachila* **sp. nov.**. Seventeen species are sorted into one of four groups associated with the species *L. fuscum*, *L. humile*, *L. iniquum*, or *L. neotropicum*, and two species are left unassigned. New generic diagnoses are provided for worker, male, and queen castes, and Shattuck's (1992a) generic descriptions of the worker, male, and queen castes are modified slightly to take into account expanded knowledge of character state variation. Worker and, where known, male and queen castes are described. Diagnoses, illustrations, and keys are supplied for worker and male castes. Discussions of variation, comparisons to similar species, and nomenclatural issues are given for each species, as well as synopses of life history traits such as queen number, colony structure, geographic distribution, nest site and habitat records, and associations with parasitoid *Pseudacteon* flies (Diptera: Phoridae).

Key words: Linepithema, taxonomy, Formicidae, systematics, monograph, natural history.

INTRODUCTION

Linepithema ants are a common but often overlooked element of the Neotropical myrmecofauna. These small, monomorphic dolichoderine ants are native to a variety of forest, grassland, and montane habitats in Central and South America and the Caribbean. The genus is widely recognized for the pestiferous Argentine ant *L. humile* (Mayr), an insect that has received considerable attention for its invasive behavior in Mediterranean climates worldwide (Roura-Pascual et al. 2004), but *Linepithema* also contains nearly twenty additional species, some of them locally abundant, about which little is known. The present paper attempts the first synthesis of the natural history and taxonomy of this rich and previously neglected group.

Natural history. The native range of *Linepithema* extends from the highlands of northern Mexico east into the Caribbean and south to northern Argentina (Fig. 109). Curiously, these ants are apparently absent from Cuba despite being abundant on both Hispaniola and Puerto Rico. They are also relatively rare in the Amazon Basin, reaching their peak abundance and diversity instead between 20° and 30° degrees south latitude (Fig. 109). The elevational range of *Linepithema* is impressive. In subtropical South America, *Linepithema* ants are found near sea level in rainforests, scrub forests, and floodplains. In the Central Andes they ascend to 4,000 meters elevation. In northern South America, Central America, and the Caribbean *Linepithema* species are more typically montane, sometimes occurring locally at high densities to the apparent exclusion of other ant species (Wild, pers. obs.). Two species, *L. iniquum* (Mayr) and the Argentine ant *L. humile*, have been carried around the world with human commerce (Fig. 108), although *L. iniquum* seems to establish only in greenhouses (Wheeler 1929, Creighton 1950).

Although *Linepithema* ants are often observed in undisturbed primary habitat, most species may also readily be found in pastures, lawns, roadsides and other disturbed habitats, suggesting that populations weather deforestation and habitat modification reasonably well. Some species, including the notorious Argentine ant, likely thrive with disturbance. Nonetheless, a few of the uncommon species have been recorded only from primary habitats and one, *L. flavescens* (Wheeler and Mann), is known only from Haiti and has not been collected since 1934.

Linepithema ants show a stereotypical nesting and feeding behavior. Mature colonies are populous, often with more than 1,000 individuals, and the worker caste is monomorphic. The trophic habits of these ants are not unusual for dolichoderines, as they are generalist scavengers and predators with strong proclivities for tending nectaries and honeydew-producing insects. Honeydew feeding is ubiquitous

1

throughout the genus and occurs both inside and outside the nest. *Linepithema* ants readily form chemical recruitment trails and can recruit in large numbers to food sources. They are commonly seen running in files on the ground and on low vegetation and are active both day and night. Some species are attacked by host-specific parasitoid *Pseudacteon* phorid flies (Orr et al. 2001, Wild, pers. obs.).

Most species are polydomous, many are also polygynous, and at least one species, the Argentine ant *L. humile*, has both multicolonial and unicolonial populations (Krieger and Keller 2003, Tsutsui and Case 2001). However, colony life history information outside of *L. humile* is sparse. Most *Linepithema* ants construct superficial nests in soil, leaf litter, rotting wood, or under stones, but at least two species, *L. iniquum* and *L. leucomelas*, are predominantly arboreal. The morphology of two poorly known species, *L. flavescens* and *L. cryptobioticum*, indicates a primarily subterranean existence although this has yet to be confirmed by field observation.

Mating behavior is unknown for most species, but there are hints of extensive variation within this genus. Queens of *L. humile* are flightless and mate in the nest (Krieger and Keller 2000), but the few other species for which observational data are known have flighted queens (Wild, pers. obs.) Interestingly, male morphology is more variable within *Linepithema* than it is among many other ant genera, suggesting that mating behavior in *Linepithema* may show extraordinary diversity. Males vary in size from much greater than the workers to somewhat smaller, with variation in the number of closed cells in the fore wing, the relative elongation of the body and wings, the degree of development of the terminalia, the degree of development of the mesosoma, and the extent to which the head and petiolar morphology approaches the female condition (Shattuck 1992c, Wild, pers. obs). Mating flights, where they have been observed, take place around dusk (Wild, pers. obs).

Taxonomic history. For most of its 140 year history, the name *Linepithema* referred only to scattered records of unusually gracile dolichoderine males from Andean South America. The Austrian myrmecologist Gustav Mayr described the genus and a new species, *L. fuscum* Mayr 1866, from several male specimens collected in the vicinity of Lima, Peru. The name subsequently lingered in obscurity for over a century, accumulating only one additional species, *L. gallardoi* Kusnezov, in 1969.

Concurrently, the heterogeneous dolichoderine genera *Hypoclinea* Mayr and *Iridomyrmex* Mayr received a steady accumulation of apparently nondescript Neotropical species, starting with *H. humilis* Mayr in 1868, described from workers. Emery (1888) soon consolidated many of these under *Iridomyrmex*, a genus originally established from Old World species. The most prolific describers of the New World *Iridomyrmex* were Forel (1885, 1907, 1908, 1911, 1912, 1913, 1914), Santschi (1916, 1919, 1923, 1929), and Wheeler (1908, 1913b, 1942, Wheeler and Mann 1914), with some additional work by Mayr (1868, 1870a), Emery (1890, 1894a, 1894b), Gallardo (1916), Borgmeier (1928), Kempf (1969), and Wilson (1985). Their work invariably consisted of isolated descriptions of new taxa and new castes. The only attempt at a synthesis was by Santschi (1929), who wrote a cursory and nearly unworkable key to an undefined "*Iridomyrmex humilis*" complex.

As the breadth of *Iridomyrmex* increased through the 20th century, some myrmecologists voiced concern that the New World forms were of different origin than the Old World *Iridomyrmex* (Brown 1958, Snelling and Hunt 1975). In a worldwide review of *Iridomyrmex*, Shattuck (1992a) concurred and removed the New World species from the genus. Nest series associating males and workers housed at Museum of Comparative Zoology in Cambridge, Massachusetts and full series from Venezuela allowed Shattuck (1992a) to connect some New World *Iridomyrmex* workers with the enigmatic *Linepithema* males and transfer the former to the latter. Shattuck (1992c) redescribed the worker, male and queen castes in considerable detail and divided the genus into two informal species groups, the Humile-group and the Fuscum-group, based on male morphology.

Since Shattuck's dolichoderine revision, *Linepithema* has received only minor taxonomic adjustments. Bolton (1995) provided the replacement name *L. inacatum* for a junior secondary homonym, *L. gallardoi* Kusnezov. Wild (2004) redescribed and rediagnosed the Argentine ant *L. humile*. Brandão et al. (1999) transferred one of two Dominican fossil species to *Technomyrmex* Mayr, and Wild and Cuezzo (2006) transferred the second fossil to *Gracilidris* Wild and Cuezzo, leaving *Linepithema* with only extant representatives.

The extant species of the newly reconstituted *Linepithema* almost certainly form a monophyletic group. Shattuck (1992c) provided a series of morphological synapomorphies for *Linepithema* within the subfamily Dolichoderinae, and a morphological cladistic analysis (Shattuck 1995) determined that *Linepithema* was likely related to a group of mostly Indo-Australian ants. This relationship is likely to hold with additional data (Wild, unpublished molecular data; Ward, unpublished molecular data), although it is worth noting that Chiotis et al. (2000) suggested a sister-taxa relationship of *Linepithema* and the thermophilic New World genus *Forelius* Emery based on mitochondrial DNA sequences. However, support for *Forelius* as sister group was weak (Chiotis et al. 2000), and mitochondrial DNA may be saturated for taxa of that age. The evolutionary history of *Linepithema* is the subject of an ongoing species-level phylogenetic study by the author and will be reported separately.

At the outset of the present revision, *Linepithema* contained 14 described extant species, 14 additional subspecies and varieties, and one fossil form (Shattuck 1994,

Bolton 1995, Brandão et al. 1999). These taxa were poorly defined and difficult to identify. Several factors have contributed to lack of taxonomic clarity. First, incoherent and inconsistent conceptions about the definition of species prior to the NeoDarwinian synthesis led to a proliferation of largely meaningless infraspecific taxa (Mayr 1942, Creighton 1950). Second, some taxonomists relied on character systems that showed unfortunate patterns of convergence. For example, *Iridomyrmex micans* Forel 1908 was described as a subspecies of *I. dispertitus* Forel 1885 based on the presence of a slight mesonotal impression, a trait that often varies within species and shows little phylogenetic pattern. Third, the described species appear to be closely related. A bewildering and often continuous morphological variation in the worker caste within and among the named species makes species delineation difficult, a situation first noted by Emery (1894c) and reiterated by Wild (2004). Fourth, *Linepithema* has suffered a paucity of taxonomic attention common to many of the small, streamlined dolichoderine genera. Lacking the spines, integument sculpture, shapes and colors considered aesthetically attractive in other tropical ant lineages, these small brownish ants have been largely passed over. As a result, the taxonomy of *Linepithema* has remained a confused scattering of isolated descriptions unbounded by any significant examination of species boundaries and with few resources for species identification.

The prominence of the pest species *L. humile* has been a mixed blessing for our understanding of *Linepithema*. On one hand, considerable resources have been directed towards understanding the physiology, ecology, genetics, and sociobiology of the Argentine ant itself (see *L. humile* species synopsis), rendering *Linepithema* one of the more thoroughly investigated of all ants and a model system for studying biological invasions. However, the prominence of this single species has lead to an unfortunate tendency of researchers to misdiagnose other species as *L. humile*. Nearly every common *Linepithema* species can be found identified as *L. humile* in major entomological collections (Wild 2004), and many literature records of the native distribution of *L. humile* (e.g., some Brazilian records in Suarez et al. 2001) are incorrect. One study on the interactions of phorid parasitoids with *Linepithema* in Brazil (Orr and Seike 1998) originally mistook *L. micans*, *L. gallardoi*, and *L. angulatum* for *L. humile*, leading to some faulty conclusions about the regulation of native Argentine ant populations by phorids that later appeared in the invasion biology literature (e.g., Chapin et al. 2000, Harris 2002).

The present work is the first comprehensive taxonomic revision of *Linepithema* ants. I rediagnose the genus, clarify species boundaries, describe several new species, and provide descriptions, diagnoses, illustrations, and keys for worker and male castes. A stable taxonomic foundation for *Linepithema* should facilitate further research on these ants.

MATERIALS AND METHODS

Specimens. I examined 8,161 worker, 301 queen, and 451 male *Linepithema* specimens collected across the global distribution of the genus. The majority of these were viewed during visits to several entomological museums and through institutional and personal loans. To supplement this material, I observed and collected *Linepithema* in the field in Argentina, Paraguay, Ecuador, Guatemala, the Dominican Republic, Jamaica, and Puerto Rico from 1996 to 2003, in a series of expeditions ranging in length from a few days to several months. These collections consisted mainly of extensive visual searches targeting *Linepithema* nests and foraging ants, sometimes augmented with baits, berlese funnels, malaise traps, and blacklights. Voucher specimens from these expeditions have been deposited in host country collections and distributed among other major entomological museums. Entomological collections cited in this study are abbreviated as follows:

ALWC- Alexander L. Wild personal collection, Tucson, Arizona, USA.
AVSC- Andrew V. Suarez personal collection, Urbana, Illinois, USA.
BMNH- British Museum of Natural History, London, UK.
CASC- California Academy of Sciences, San Francisco, California, USA.
IFML- Instituto Fundación Miguel Lillo, Tucumán Argentina.
INBP- Museo Nacional de la Historia Natural del Paraguay, San Lorenzo, Paraguay.
JTLC- John T. Longino personal collection, Evergreen, Washington, USA.
LACM- Natural History Museum of Los Angeles County, Los Angeles, California, USA.
MACN- Museo Argentina de Ciencias Naturales, Buenos Aires, Argentina.
MCSN- Museo Civico de Historia Natural 'Giacomo Doria', Genoa, Italy.
MCZC- Museum of Comparative Zoology, Cambridge, Massachusetts, USA.
MHNG- Muséum d'Histoire Naturelle, Geneva, Switzerland.
MZSP- Museu de Zoologia da Universidade de São Paulo, São Paulo, Brazil.
NHMB- Naturhistorisches Museum, Basel, Switzerland.
NHMW- Naturhistorisches Museum Wien, Vienna, Austria.
PSWC- Philip S. Ward personal collection, Davis, California, USA.
QCAZ- Museo de Zoología de la Pontificia Universidad Católica del Ecuador, Quito, Ecuador.
UCDC- R. M. Bohart Museum of Entomology, Davis, California, USA.
USNM- National Museum of Natural History, Washington, D.C., USA.
WPMC- William P. MacKay personal collection, El Paso, Texas, USA.

Many of the older names in *Linepithema* had been described from long series of *de facto* syntype specimens, and some of these series inadvertently contain specimens of ants from different genera or even different subfamilies. In order to preserve nomenclatural stability, I have designated and clearly labeled lectotypes from each of these series for the following valid and invalid taxa: *L. angulatum*, *L. arrogans*, *L. bicolor*, *L. breviscapa*, *L. dispertitum*, *L. dominicense*, *L. flavescens*, *L. fuscecens*, *L. fuscum*, *L. impotens*, *L. keiteli*, *L. leucomelas*, *L. melleum*, *L. micans*, *L. nigellum*, *L. platense*, *L. pordescens*, *L. riograndense*, *L. scotti*, *L. subfasciatum*, and *L. succineum*.

Most of the new species described in the present work were represented by multiple collections. Type series for each new species were selected from the available material using two main criteria. First, nest series containing males, queens, and larger numbers of workers were chosen over smaller series and series with fewer castes. Second, series collected from reserves, parks, or other protected habitats were given preference over collections from areas with less protection. Holotype specimens of new species have been returned to institutions in their country of origin, and paratypes from the same series as the holotypes have been distributed broadly among major myrmecological collections.

Type series of many new species are described from Paraguay even though most of these species also occur in Brazil and elsewhere. The reason for this choice is largely bureaucratic. At the time of field research, the collection and specimen export permitting process for Paraguay, while involving a few weeks' processing time, was faster and less difficult than the comparable process for Brazil. Consequently, targeted collecting in Paraguay made available long nest series of fresh material.

Morphological analysis. Most observations were made at 50x on a Wild stereo microscope. I conducted morphometric measurements on a subset of male (n = 79), queen (n = 65), and worker specimens (n = 526). A majority of measurements were taken using a dual-axis Nikon stage micrometer with a precision of 0.001 mm, but measurements at IFML and MZSP employed an ocular micrometer with a precision of 0.01 mm. I repeated measurements on several specimens using both optical and stage micrometers to confirm that measurements were consistent between systems. Because of differences in measurement precision between systems, and because of a small but inevitable amount of measurement error, I report measurements here to 0.01 mm.

I employed a number of standard morphometric characters. Head measurements are given with the head in full face view, with the anterior clypeal margin and the posterior border of the head in the same focal plane. I consider ant heads to be prognathous, such that the clypeus is anterior and the frontal area is dorsal.

HL - Head length. In full face view, the midline distance from the level of the maximum posterior projection of the posterior margin of the head to the level of the most anterior projection of the anterior clypeal margin. In males, I consider the posterior margin of the head as the vertex between, and not including, the ocelli.

HW - Head width. In full face view, the maximum width of the head posterior to the compound eyes.

MFC - Minimum frontal carinal width. In full face view, the minimum distance between the frontal carinae.

SL - Antennal scape length. Measured from the apex of the first antennal segment to the base, exclusive of the radicle.

FL - Profemur length. In posterior view, measured along the longitudinal axis from the apex to the junction with the trochanter.

LHT - Metatibial length. In dorsal view, measured along the longitudinal axis from the apex to the level of the lateral condyles, excluding the medial proximal condyle.

PW - Pronotal width. In dorsal view, the maximum width of the pronotum measured from the lateral margins.

WL - Wing length. In males and queens, the maximum distance between the base of the sclerotized wing veins to the distal margin of the wing.

MML - Maximum mesosomal length. In males and queens, the distance from the maximum anterior projection of the mesosoma to the maximum posterior projection of the propodeum. In males with a well developed mesosoma the anterior projection of the mesosoma is often formed by a swollen mesoscutum, and the posterior projection is formed by a rearward projection of the propodeal dorsum above the petiole. MML in workers was not taken because the flexible articulation between the pronotum and the mesonotum introduces substantial variation in this measurement.

EL - Eye length. In full face view, the length of the compound eye along the longitudinal axis.

EW - Eye width. With eye held in focal plane facing the viewer, the maximum transverse width of the compound eye.

ES - Eye size. 100*EL*EW.

SI - Scape index. 100*SL/HL.

CI - Cephalic index. 100*HW/HL.

CDI - Carinal distance index. 100*MFC/HW.

OI - Ocular index. In workers, 10*ES/HL (= 1000*EL*EW/HL). In males and queens, 100*EL/HL.

WI - Wing index. In males and queens, 10*WL/MML.

FI - Femoral index. In males and queens, 100*FL/MML.

In addition to morphometric characters, I examined a suite of morphological characters commonly used in ant systematics. These characters include pilosity, pubescence, body color, wing venation, shape of the head, shape of the mesosoma and associated sclerites, shape of the petiole, male genital morphology, and maxillary palp morphology. Anatomical terms are illustrated for workers in Figures 1-3, venational characters in Figures 47-48, and male genital characters in Figures 49-51 and 53. Additional terminological guidance may be found in the more extensive glossary provided by Bolton (1994).

Several characters of particular importance to the taxonomy of *Linepithema* merit additional explanation.

Pubescence. In most specimens the pubescence forms dense, velvety mats comprised of small, hair-like cuticular projections on at least part of the dorsal surface of the ant. Proper examination of the pubescence requires diffuse lighting at sufficient magnification. If a spot light source is the only light available (as in many fiber-optic lighting systems), the light should be reflected off a white card held close to the specimen instead of pointed directly at the specimen.

The density and extent of the pubescence on various sclerites can vary in taxonomically informative ways, especially among workers. In particular, the pubescence on the mesopleuron and on the metapleuron can be absent (Fig. 4), dense (Fig. 5), or fading to sparse or absent anteroventrally (Fig. 6). The first couplet in the worker key relies on the distance between the appressed pubescent hairs on the metapleuron. In most species the metapleuron is reliably either densely pubescent with tightly spaced hairs or nearly entirely devoid of hairs. This character is ambiguous in some populations of a few common species, and the key has been designed so that these species key out on both sides of the couplet.

Propodeal shape. The propodeum is the posterior dorsal sclerite on the mesosoma. The shape of the propodeum is variable among *Linepithema* workers and is reliably diagnostic of some species or species groups. Unfortunately the shape variation is complex and difficult to describe succinctly, so I have avoided making heavy use of propodeal characters in the keys.

Species in the Humile-group have a propodeal profile with a characteristic angular shape (as in Figs. 31, 35, 37), the posterior face is "broken" at about the level of the propodeal spiracle into two nearly flat margins of differing slope. This shape gives the propodeum of most Humile-group species an anteriorly-inclined appearance. The metapleural bulla in these species also tends to be reduced and nearly flush with the posterior edge of the propodeum.

Species in the Neotropicum-group have a relatively low, rounded propodeum (Figs. 27, 29), bearing a long dorsal and a short posterior face that meet each other at a curve or an indistinct, rounded angle. These species also have a metanotal groove that is only weakly or not at all impressed.

Workers of some species are diagnosable entirely through propodeal configuration. The propodeum of *L. aztecoides* is instantly recognizable as it is unusually low (Fig. 43) with a posterodorsal concavity for the reception of the petiole when the gaster is held over the body. In this configuration the posterolateral corners of the propodeum are formed by the propodeal spiracles. Most populations of *L. iniquum* have a globular propodeum that is set off from the mesonotum by a strongly impressed metanotal groove (Fig. 23). *L. pulex* workers bear a propodeum with a somewhat concave posterior face ventrally (Fig. 45).

In males, propodeal shape can be diagnostic for species groups. The Humile-group shows an enlarged propodeum with a greatly concave posterior margin in lateral view such that the most posterior projection of the propodeum overhangs the petiole (Figs. 52, 54). Males in the Iniquum (Fig. 53) and Fuscum (Fig. 51) groups normally have a smaller, more rounded propodeum with a straight to convex posterior face. Males of other species usually have the posterior propodeal margin straight to weakly convex and not overhanging the petiole.

Mesonotal shape. The mesonotum is the middle dorsal sclerite on the mesosoma of worker ants. The mesonotal shape in profile is rather variable, both within and among species, and is consequently only of limited taxonomic utility. The shape may be robust, with a strongly convex profile (Fig. 15), straight (Figs. 25, 29) or with varying degrees of mesonotal impression (Figs. 21, 23). In the arboreal species *L. iniquum*, the impression is always present and usually deep enough that the maximum width of the mesosoma at the point of constriction is narrower than the maximum diameter of the fore coxa.

Maxillary palps. The maxillary palps in *Linepithema* are 6-segmented. The overall length of the palps as well as the relative lengths of the component segments varies among workers of different species. The palps are best examined in slide mounts of the mouthparts, but can often be seen in pinned specimens. In specimens where the palps are greater than ½ head length (as in Fig. 43), segments 4 and 5 (counting from the base) are usually subequal in length to the adjoining segments, and the palps when laid out along the venter of the head surpass the level of the posterior extension of the compound eyes. In specimens where the palps are shorter than ½ head length (as in Fig. 31), segments 4 and 5 are usually reduced relative to the adjoining segments, and the palps when laid out along the venter of the head do not surpass the level of the posterior extension of the compound eyes. Although relative palp length can be diagnostically useful, the worker key here does not rely on palp characters because the palps can be difficult to see and are sensitive to differences in specimen preparation.

Wing venation. The venational terminology used here is that of Brown and Nutting (1950). *Linepithema* males show either of two distinct venational patterns. The pattern of Fuscum-group males is similar to that of queens, where the substigmal region of these ants is divided by a longitudinal vein of the radial system, Rs, so that there are two elongate submarginal cells (Fig. 47, "smc1, smc2"). All other *Linepithema* males have only one submarginal cell (Fig. 48, "smc1").

Queens show variation in the relative lengths of the Rs+M and the M.f2 veins that border the submarginal cells on the proximal side (as in Fig. 47 "Rs+M, M.f2"). Rs+M may be longer, subequal, or shorter in length than M.f2.

Volsella. The volsella is a male reproductive appendage that sits interior of the parameres and external of the aedeagus. In most *Linepithema* species the volsella is short (as in Fig. 50), partly obscured by the parameres in undissected specimens (as in Figs. 52-54), and shows little differentiation among species. In the highly-modified Fuscum-group, the volsella shows strong and reliable species-specific differences (Figs. 56, 58, 60, 62, 64). These males have strongly extruding terminalia such that the volsella is nearly always plainly visible without dissection (as in Fig. 51).

The volsella consists of a large basal lobe that bears an elongate dorsal and distal projection called the digitus, often a ventral distal tooth (Figs. 49–50, "*vp*"), and sometimes a distal cuspis (Fig. 50, "*c*"). In all *Linepithema* males, the digitus is downturned distally such that the digitus consists of a proximal arm that is somewhat parallel to the body of the volsella and a distal arm oriented ventrally. The relative lengths of the distal and proximal arms varies informatively between species, as does the width of the base of the proximal arm.

Species delimitation. I employ the view that species are aggregates of interbreeding or potentially interbreeding populations (the Biological Species Concept, Mayr 1942). Although resources were not available in the present study to directly examine gene flow, species boundaries can be inferred indirectly through morphological and geographic data (Coyne and Orr 2004). Specifically, character states within biological species are likely to be continuous, while character states may be expected to diverge in the absence of gene flow, leaving a distinct gap. Consequently, consistent gaps in morphology between sympatric populations that have opportunity for interbreeding can be taken as a proxy for reduced gene flow and for species boundaries. In the present paper, sympatric character state discontinuities are the primary criteria for recognizing species. Some morphometric examples are given in Figures 77–99.

Unfortunately, not all populations occur in sympatry, either locally or regionally. Allopatric populations are widely recognized as problematic for the Biological Species Concept (Wheeler and Meier 2000). In cases of extreme morphological and ecological disparity, such as the rare chaco species *L. cryptobioticum*, the establishment of a new species from an isolated population is straightforward. However, most cases of allopatry present a range of variation that is considerably more difficult. I describe the observed variation in the discussion section of each species synopsis. Where the evidence is insufficient to either resolve a population into a pre-existing species or to elevate it to species status, I have taken a nomenclaturally conservative stance and treated ambiguous populations as conspecific with known species. While this approach surely obscures some real biological diversity, it prevents a proliferation of names that will later require synonymy.

The final product of species delineation in this paper is not strictly a reflection of biological species, then, but rather an inference of biological species weighted by utilitarian concerns in proportion to the degree of uncertainty in the inference. The resulting species serve as hypotheses that may be tested with additional data. It is my hope, in fact, that this revision will encourage the more detailed research in *Linepithema* that eventually renders the present study obsolete.

Finally, it bears noting that concurrent with this taxonomic revision I have been reconstructing a species-level phylogeny for *Linepithema* based on DNA sequence data at 4 loci. These molecular results will be reported elsewhere. The insights gained through the molecular phylogeny, though carrying some heuristic purpose in the early stages of species delineation, ultimately have not been used to define species boundaries out of concern that taxon sampling may disproportionately influence results. As may be expected from evolutionary processes, some widespread species are likely paraphyletic (Funk and Omland 2003). These are noted in the discussion section of each species synopsis.

Geography. Distribution data were taken from specimens examined during the course of this study and do not include literature records. Maps were drawn in the shareware program Versamap (www.versamap.com) on a Windows PC computer platform using coordinates provided on specimen labels or inferred from maps and gazetteers for those specimens without coordinate data. A number of older specimens did not have sufficiently detailed labels to infer exact coordinates (e.g., "Colombia") and were excluded from mapping, although the information about these specimens appears in the material examined section of the species synopses.

In the material examined sections, locality data are presented as they appear on the specimen labels and may include some older or out of date names for administrative divisions, particularly in Brazil. Latitude and longitude coordinates are given only when such data appeared on the original specimen label.

Illustrations. Line drawings from preserved specimens were traced by hand using a Wild camera lucida mounted on a Wild M5 stereomicroscope at 100x, inked, scanned at 800dpi, and cleaned in Adobe Photoshop (www.adobe.com/products/photoshop/main.html) on a Windows PC platform. A small number of figures (Figs 4–6) showing pubescence differences are fully-focused montage images created using a Leica MZ16 stereomicroscope at 115x with the Syncroscopy Auto-Montage software package (www.syncroscopy.com).

TAXONOMIC RESULTS

Genus *Linepithema* Mayr

Linepithema Mayr 1866: 496. Type species: *Linepithema fuscum* Mayr, by monotypy.

Worker diagnosis (key characters in bold). Small dolichoderine ants (HW 0.42–0.80) with a monomorphic worker caste. Compound eyes comprising 17–110 ommatidia, centered anterior of midline of head in full face view, not touching lateral margins; **mandible with dentition consisting of an elongate apical tooth and a smaller subapical tooth followed by a series of 3–4 small teeth alternating with denticles** (as in Fig. 3), masticatory and inner margins meeting at a curve armed with 1–3 teeth or denticles; **anteromedial clypeal margin with a broad, shallow concavity**; palp formula 6:4; mesosoma lacking spines or teeth; propodeum in lateral view depressed below level of mesonotum; fourth gastric sternite keel-shaped posteriorly; pilosity moderate to reduced, head lacking standing setae along posterolateral corners and pronotum bearing fewer than 10 standing setae.

Queen diagnosis. Mandible with dentition consisting of an elongate apical tooth and a smaller subapical tooth followed by a series of 3–5 small teeth alternating with denticles (as in Fig. 3), masticatory and inner margins meeting at a curve armed with 1–3 teeth or denticles; anteromedial clypeal margin with a broad, shallow concavity; palp formula 6:4; axilla with a medial suture; mesoscutum covered with a dense, fine pubescence; venter of petiole with a slight lobe.

Male diagnosis. (Excluding some worker-like males in populations of *L. dispertitum*; see species synopses for description and discussion of variation in *L. dispertitum*). Antennal scape shorter than third antennal segment; lateral ocelli emerging above posterior cephalic margin in full face view; anteromedial clypeal margin broadly convex; mandibles with a distinct masticatory margin bearing at least 4 teeth or denticles, sometimes approaching worker dentition; mesoscutum covered with dense, fine pubescence; petiolar scale not inclined anteriorly, instead with scale straight, inclined posteriorly, or present as a low node; forewings with 1–2 closed submarginal cells; hind wings with 2 closed cells; volsella with digitus narrow and sharply downturned distally.

Descriptions: The detailed generic descriptions of worker, queen, and male castes provided by Shattuck (1992c: 114–118) are generally valid and will not be altered substantially here. However, the following minor corrections are necessary [with Shattuck's (1992c) original character state descriptions in brackets]:

Worker: Third maxillary palp segment subequal in length to about 1/3 longer than segment 4 [was, "third maxillary palp segment subequal in length to segment 4."]; erect pronotal hairs 0–8 (most commonly 0 or 2) [was, "pronotal hairs 0–6 (generally 2)"]; propodeal spiracle lateral and usually ventral of the propodeal dorsum, rarely forming posterodorsal corners of the propodeum (*L. aztecoides*) [was, "propodeal spiracle lateral and ventral of the propodeal dorsum."]; gastral compression usually absent, rarely dorso-ventrally compressed (*L. aztecoides*) [was, "gastral compression absent"].

Queen: Vertex slightly convex to concave [was, "Vertex flat to weakly concave"]; scape short, surpassing vertex by less than 1/3 scape length [was, "by less than one-half scape length"]; third maxillary palp segment subequal in length to about 1/3 longer than segment 4 [was, "third maxillary palp segment subequal in length to segment 4."]; erect mesoscutal hairs 2–35 [was, "2–25"]; forewing with 2 submarginal cells and one discoidal cell [was, "forewing with 2 cubital cells and 1 discoidal cell", the closed cells referred to by Shattuck are in the radial sector, not the cubital sector (Ward and Downie 2005)].

Male: Apical tooth of mandible varying from shorter to much longer than subapical tooth [was, "subequal in length to, to slightly longer than, the subapical tooth"]; forewing with 1–2 submarginal cells and one discoidal cell [was, "forewing with 1–2 cubital cells and 1 discoidal cell", the closed cells referred to by Shattuck are in the radial sector, not the cubital sector (Ward and Downie 2005)].

Distribution. Native from northern Mexico and the Caribbean south to northern Argentina, with introduced populations worldwide.

Discussion. The composition and the definition of *Linepithema* established by Shattuck (1992a, 1992c) are largely upheld in the present revision. A comprehensive review of character states across the genus warrants the minor expansions to the generic description listed above, as well as a modification to Shattuck's (1992c) informal species groups.

Shattuck's division of *Linepithema* into two species groups based on male morphology is incomplete. Although his Fuscum-group is undoubtedly monophyletic, his Humile-group lacks synapomorphies (Shattuck 1992c) and is probably paraphyletic (Wild, unpublished molecular data). Here I divide *Linepithema* into four potentially monophyletic groups associated with the species *L. fuscum*, *L. humile*, *L. iniquum*, and *L. neotropicum*, with two species left unassigned. This organization of *Linepithema* into species groups is intended to facilitate communication, but it should be noted that these groups are informal and not regulated by the codes of nomenclature established by the ICZN. Diagnostic character states for each group are given in Table 1.

The most reliable way to diagnose *Linepithema* workers to genus is the combination of distinctive mandibular dentition (Fig. 3) and the anteromedial clypeal concavity. Additionally, keys in Shattuck (1992) and Bolton (1994) serve to identify workers to genus. The rare, recently described ant *Gracilidris* Wild & Cuezzo (2006) will also key to *Linepithema*, but workers of that genus have the compound eyes higher and more lateral on the head and lack the concave anteromedial clypeal margin of *Linepithema*. The dolichoderine key of Palacio and Fernandez (2003) will work for some species, but cannot be recommended as *L. aztecoides* and any specimen with a robust mesonotum, such as some *L. piliferum*, may incorrectly key to *Azteca*. Santschi's (1929) key to the workers of the "*Iridomyrmex humilis*" group is out of date and should not be used.

Table 1. Male and worker characteristics useful for diagnosis of *Linepithema* species groups

| species group | **Male traits** | | | | | **Worker Traits** | | | | |
	Number of submarginal cells in forewing	Relative length of gonostylus	Relative length of volsella	Presence of cuspis	Shape of posterior propodeal margin	Metapleural pubescence	Propodeal shape	Maxillary palp length	Cephalic pilosity, excluding clypeus	Pre-sutural clypeal groove
Fuscum	*2	*long	*long	*absent	*convex	sparse to absent	variable	variable	variable, 0-9 standing setae	*usually well-developed
Humile	1	short	short	present	*strongly concave and overhanging petiole	*dense	*inclined forward, posterior profile broken at spiracle	subequal to or shorter than 1/2 HL	variable, 0-6 standing setae, often absent	weakly developed or absent
Iniquum	1	short	short	present	*convex	sparse to absent	variable	longer than 1/2 HL	variable, 0-13 standing setae, often extensive	weakly developed or absent
Neotropicum	1	short	short	present	weakly concave	moderate to sparse	*low, relatively rounded, dorsal face longer than posterior face	longer than 1/2 HL	absent	weakly developed or absent
incertae sedis *L. aztecoides*	n/a; males unknown	n/a	n/a	n/a	n/a	sparse to absent	* strongly compressed dorsally, concave between spiracles	longer than 1/2 HL	0-4	weakly developed or absent
L. pulex	1	short	short	present	weakly concave	sparse to absent	prominent	subequal to or shorter than HL	0-2	moderately developed

*putative synapomorphy

Taxonomic synopsis of *Linepithema* species

Fuscum-group

    angulatum (Emery 1894a) **stat. nov.** Costa Rica to central Brazil.

        = *pordescens* (Wheeler 1942) **syn. nov.**

    cryptobioticum **sp. nov.** Paraguay.

    flavescens (Wheeler & Mann 1914) **stat. nov.** Hispaniola.

    fuscum Mayr 1866 Ecuador and Peru.

    keiteli (Forel 1907) Hispaniola.

        = *subfaciatum* (Wheeler & Mann 1914) **syn. nov.**

    piliferum (Mayr 1870a) Costa Rica to Ecuador and Venezuela.

    tsachila **sp. nov.** Colombia and Ecuador.

Humile-group

    anathema **sp. nov.** Brazil.

    gallardoi (Brèthes 1914) Colombia to Argentina.

        = *breviscapa* (Santschi 1929) **syn. nov.**

        = *impotens* (Santschi 1923) **syn. nov.**

    humile (Mayr 1868) Native to northern Argentina, Paraguay, and southern Brazil; introduced worldwide.

        = *arrogans* (Chopard 1921) [syn. by Wild {2004}]

        = *riograndense* (Borgmeier 1928) [syn. by Wild {2004}]

    micans (Forel 1908) **stat. nov.** Brazil to Argentina.

        = *platense* (Forel 1912) **syn. nov.**

        = *scotti* (Santschi 1919) **syn. nov.**

    oblongum (Santschi 1929) Bolivia and Argentina.

Iniquum-group

    dispertitum (Forel 1885) Mexico to Panamá, Hispaniola.

    iniquum (Mayr 1870a) Mexico to southern Brazil, Caribbean.

        = *bicolor* (Forel 1912) **syn. nov.**

        = *dominicense* (Wheeler 1913b) **syn. nov.**

        = *fuscescens* (Wheeler 1908) **syn. nov.**

        = *melleum* (Wheeler 1908) **syn. nov.**

        = *nigellum* (Emery 1890) **syn. nov.**

        = *succineum* (Forel 1908) **syn. nov.**

    leucomelas (Emery 1894b) Southeastern Brazil.

        = *aspidocoptum* (Kempf 1969) **syn. nov.**

Neotropicum-group
> *neotropicum* **sp. nov.** Costa Rica to southeastern Brazil.
> *cerradense* **sp. nov.** Bolivia, Brazil, and Paraguay.

unassigned to group
> *aztecoides* **sp. nov.** Brazil and Paraguay.
> *pulex* **sp. nov.** (first available use of *pulex* Santschi 1923) Brazil, Paraguay, and Argentina.

incertae sedis
> *inacatum* Bolton 1995 (replacement name for *L. gallardoi* Kusnezov 1969, junior secondary homonym, type specimens not located. Possibly a junior synonym of *L. angulatum* Emery 1894a)

Names excluded from *Linepithema*

humilioides (Wilson 1985)

Iridomyrmex humilioides Wilson 1985: 33 (W). Holotype worker, Dominican Amber [MCZC, examined].
Linepithema humilioides (Wilson) Shattuck 1992a: 16. First combination in *Linepithema*.
Gracilidris humilioides (Wilson) Wild and Cuezzo 2006: 66. First combination in *Gracilidris*.

This species is known from a single fossil worker specimen, briefly described by Wilson (1985) and placed in the genus *Iridomyrmex*. At the time, *Iridomyrmex* had a broad definition encompassing many old world and new world forms. When Shattuck (1992) re-defined *Iridomyrmex* and moved most of the new world species into *Linepithema*, he transferred *Gracilidris humilioides* as well, even though *G. humilioides* lacks many of the apomorphies that define extant *Linepithema*. Wild and Cuezzo (2006) transferred this species to the extant genus *Gracilidris*, and that arrangement is upheld in the present study.

Key to the species of *Linepithema* based on workers

1. Metapleuron anterior of metapleural gland orifice with a central area devoid of pubescence; if some appressed hairs present, then distance between hairs is greater than length of hairs (Fig. 4); those specimens ambiguous or intermediate in this character (similar to Fig. 6) should key out through both couplets 2

Metapleuron anterior of metapleural gland orifice with many appressed hairs on surface, distance between hairs less than length of hairs, hairs often densely spaced but sometimes fading to sparse ventrally (Figs. 5-6); those specimens ambiguous or intermediate in this character (similar to Fig. 6) should key out through both couplets 16

2. (1) Antennal scapes short (SI < 86), in full face view scapes in repose do not surpass posterior margin; eyes small, with fewer than 40 ommatidia; body color pale yellow to light brown, never medium brown to black 3

Antennal scapes longer (SI > 87), in full face view scapes in repose surpass posterior margin; eye size variable, usually with more than 40 ommatidia (35–50 in *L. fuscum*); body color variable 4

3. (2) Eyes with < 25 ommatidia; antennal scapes very short, SL < 0.45; frontal carinae relatively widely spaced, CDI > 30; petiolar scale in lateral view relatively tall, exceeding level of propodeal spiracle (Fig. 9); propodeum only slightly depressed below level of promesonotum; Paraguay .... *L. cryptobioticum* (pg. 41)

Eyes with > 25 ommatidia; antennal scapes longer, SL > 0.50; frontal carinae relatively narrowly spaced, CDI < 29; petiolar scale in lateral view not exceeding level of propodeal spiracle (Fig. 11); propodeum depressed well below level of promesonotum; Hispaniola .... *L. flavescens* (pg. 49)

4. (2) Head and mesosoma dorso-ventrally flattened (Fig. 43); propodeal spiracles produced into prominent posterolateral corners of the propodeum; petiolar node low and blunt; Brazil and Paraguay .... *L. aztecoides* (pg. 34)

Head and mesosoma not dorso-ventrally flattened; propodeal spiracles not prominent; petiolar node not low, produced instead as an anteriorly-inclined scale 5

5. (4) Mesonotum slender with a strong, step-like medial impression (Fig. 23); cephalic dorsum in undamaged specimens with more than 5 standing setae; Central America and the Caribbean to southern Brazil .... *L. iniquum* (pg. 67)

Mesonotum more robust, usually without a strong, step-like medial impression; if medial impression present (some populations of *L. dispertitum* and *L. angulatum*), then cephalic dorsum with fewer than 5 standing setae; setae on cephalic dorsum otherwise variable 6

6. (5) With the following combination of character states: propodeum in lateral view relatively low, with posterior face distinctly shorter than dorsal face (Figs. 27, 29); metanotal groove weakly or not at all impressed; metapleuron with at least a few scattered appressed setae anterior of metapleural gland opening (as in Fig. 6) 24

Either propodeum in lateral view with posterior face longer than or subequal in length to dorsal face (as in Fig. 21); or with mesonotal groove moderately to strongly impressed (as in Fig. 21); or with metapleuron completely devoid of appressed setae anterior of metapleural gland opening (as in Fig. 4) 7

7. (6) Pronotum lacking erect setae; body color medium brown to piceous; Mexico to Costa Rica and Hispaniola .... *L. dispertitum*, part (pg. 43)

Pronotum bearing erect setae (may be missing in damaged specimens); color variable 8

8. (7) Mexico; color medium brown to piceous .... *L. dispertitum*, part (pg. 43)

South America, the Caribbean, and Central America south of Mexico; color variable 9

9. (8) Sculpture on dorsum of head superficial, limited to small punctures associated with pubescence, surface shining through pubescence; head immediately laterad of eyes with pubescence sparse; head and body relatively robust (CI > 92); gastric tergite 1 with > 10 erect to subdecumbent setae; Hispaniola .... *L. keiteli* (pg. 76)

Sculpture on dorsum of head densely and finely punctate/pubescent, surface appearing dull or opaque; head immediately laterad of eyes with pubescence moderate to dense; habitus variable; pilosity variable; Central and South American mainland 10

10. (9) Pubescence moderate to sparse on gastric tergites 3–4 (= abdominal tergites 5–6); mesonotum relatively angular; metanotal groove usually deeply impressed (Fig.

45); small species (HL < 0.65); body color testaceous to reddish-brown; Atlantic forest regions of Brazil, Paraguay and Argentina .... *L. pulex* (pg. 105)

Pubescence dense on gastric tergites 3–4; other characters variable; Costa Rica south to the Pantanal 11

11. (10) Cephalic dorsum bearing standing setae near vertex 12

Cephalic dorsum lacking standing setae near vertex 15

12. (11) Central America 13

South America 14

13. (12) Antennal scapes longer, SI > 99, in full face view scapes in repose surpass posterior margin by a length greater than or equal to that of first funicular segment. .... *L. piliferum*, part (pg. 101)

Antennal scapes shorter, SI < 97, in full face view scapes in repose surpass posterior margin by a length less than that of first funicular segment .... *L. angulatum*, part (pg. 29)

14. (12) Head in full face view with posterior margin deeply concave (Fig. 20); head usually reaching widest point near level of compound eyes; antennal scapes relatively short (SI < 97), in full face view scapes in repose surpass posterior margin by a length less than length of first funicular segment; Ecuador and Colombia .... *L. tsachila*, part (pg. 109)

Head in full face view with posterior margin straight or only slightly concave (Fig. 18); head usually reaching widest point posterior of compound eyes; antennal scapes relatively long (SI < 99), in full face view scapes in repose surpass posterior margin by a length greater than or equal to length of first funicular segment; northwestern South America to Costa Rica .... *L. piliferum*, part (pg. 101)

15. (11) Head width < 0.53; eyes relatively small, with 35–50 ommatidia; metanotal groove not impressed or only weakly impressed (Fig. 13); body color medium to dark brown, never testaceous; Andean South America .... *L. fuscum* (pg. 51)

Head width > 0.53; eyes of moderate size, with 45–70 ommatidia; metanotal groove usually strongly impressed (Fig. 7); body color testaceous to medium brown; Costa Rica to southern Brazil .... *L. angulatum*, part (pg. 29)

16. (1) Mesopleural pubescence dense throughout (Fig. 5) 17

Mesopleural pubescence fading to moderate, sparse, or absent anteroventrally (Figs. 4, 6) 20

17. (16) Pubescence on gastric tergites 2–4 (= abdominal tergites 4–6) moderate to sparse on most specimens in a series, gastric dorsum strongly shining; elongate ants with long antennal scapes (SI 120–139); Andean Bolivia and Argentina .... *L. oblongum* (pg. 97)

Pubescence dense on all gastric tergites; antennal scape length variable but SI never exceeding 130 18

18. (17) With the following combination of characters: pronotum lacking standing setae; 1st and usually 2nd gastric tergite (= abdominal tergite 4) lacking standing setae; eyes relatively large, with more than 80 ommatidia, OI > 30; cosmopolitan .... *L. humile* (pg. 61)

Without the preceding combination of characters: 2nd gastric tergite bearing standing setae; pronotal setae present or absent; eye size variable 19

19. (18) Antennal scapes long (SI 119–126); head in full face view narrow (CI 80–85); pronotum lacking standing setae; Brazil .... *L. anathema* (pg. 27)

Antennal scapes of moderate length (SI 97–110); head in full face view moderate to broad (CI 84–95); pronotum with setae present or absent; Argentina, Paraguay, and Brazil .... *L. micans* (pg. 85)

20. (16) Strikingly bicolored, head dark brown and contrasting with pale whitish yellow on all or part of the mesosoma and appendages; antennal scapes long (SI 106–125); southeastern Brazil .... *L. leucomelas* (pg. 81)

Coloration not as above; if bicolored, contrast is between shades of light reddish brown and dark brown; scape length variable 21

21. (20) Pronotum relatively pilose, bearing 2–8 standing setae of varying length, the longest seta noticeably exceeds the length of the compound eye (Fig. 19); head in full face view with posterior margin deeply concave (Fig. 20); antennal scapes relatively short (SI < 97), in full face view scapes in repose surpass posterior margin by a length less than that of first funicular segment; Ecuador and Colombia .... *L. tsachila*, part (pg. 109)

Pronotum with standing setae noticeably shorter than length of the compound eye (as in Fig. 37) or absent; head shape variable; antennal scape length variable 22

22. (21) Maxillary palps shorter than ½ head length, palps when fully extended along cephalic venter do not exceed level of posterior extension of compound eye; posterior margin of propodeum in profile relatively straight above the level of propodeal spiracle and inclined anteriorly, meeting dorsal face at an angle (Fig. 35); mesonotal dorsum usually angular with distinct long dorsal and short posterior faces (Fig. 35); antennal scapes in repose exceeding posterior margin of head by a length less than or subequal to that of first funicular segment (SI 89–107, usually < 100); Argentina to Venezuela .... *L. gallardoi* (pg. 55)

Maxillary palps longer than ½ head length, palps at full extension along cephalic venter exceed level of posterior extension of compound eye; other characters variable 23

23. (22) Propodeum in lateral view with posterior face longer than or subequal in length to dorsal face (Fig. 21); mesonotal groove usually moderately to strongly impressed (Fig. 21); body size larger (HL > 0.70); color dark brown to piceous; Mexico, Central America and Hispaniola .... *L. dispertitum*, part (pg. 43)

Propodeum in lateral view with posterior face distinctly shorter than dorsal face (Figs. 27, 29); mesonotal groove weakly or not at all impressed; body size smaller (HL < 0.70); color variable; Costa Rica to southern Brazil 24

24. (6, 23) Mesosoma in profile with propodeum depressed only slightly below level of mesonotum (Fig. 27); antennal scapes relatively short (SL 0.44–0.54), in repose surpassing posterior margin of the head by a length less than that of the first funicular segment; body color testaceous to medium brown, never dark brown or piceous; Bolivia and Paraguay to eastern Brazil .... *L. cerradense* (pg. 37)

Mesosoma in profile with propodeum depressed well below level of mesonotum (Fig. 29); antennal scapes somewhat longer (SL 0.50–0.64), surpassing posterior margin of the head by a length subequal to that of the first funicular segment; body color variable, often dark brown; Costa Rica to southern Brazil ... *L. neotropicum* (pg. 91)

Key to the species of *Linepithema* based on males

Males of *L. anathema*, *L. aztecoides*, *L. cryptobioticum*, and *L. flavescens* are not known and are not included in key.

1. Forewing with one submarginal cell (Fig. 48); gonostylus produced as a robust lobe (Figs. 52-54) 2

Forewing with two submarginal cells (Fig. 47); gonostylus filamentous (Fig. 51, Fuscum-group) 11

2. (1) Propodeum with a strongly concave posterior face; propodeum produced posterodorsally and overhanging petiole (Figs. 52, 54); petiolar scale in lateral view tall with a relatively sharp crest (Humile-group) 3

Propodeum with posterior face convex, straight, or weakly concave; propodeum not strongly overhanging petiole; petiolar scale variable 6

3. (2) Mesosoma greatly enlarged (MML > 1.40); pronotum produced strongly forward and overhanging head (Fig. 54); mesosoma noticeably longer than metasoma; cosmopolitan .... *L. humile* (pg. 61)

Mesosoma not so enlarged (MML < 1.20); pronotum only weakly or not at all overhanging head (as in Fig. 52); mesosoma subequal in length to metasoma 4

4. (3) Legs relatively long (FI > 56); compound eyes small (OI 40–44); Andean Bolivia and Argentina .... *L. oblongum* (pg. 97)

Legs shorter (FI < 56); compound eyes of variable size (OI 40–49, usually > 45); South America 5

5. (4) Either pubescence moderate to sparse on head posterodorsal to compound eye with surface shining, or collected in South America from Peru northward .... *L. gallardoi* (pg. 55)

Pubescence moderate to dense on head posterodorsal to compound eye, and collected in South America from Bolivia southward .... *L. micans* (pg. 85)

6. (2) First antennal segment relatively long (SI > 31); eye separated from posterolateral clypeal margin by a distance greater than or equal to width of antennal scape (Figs. 70-72); Central America and the Caribbean ...... *L. dispertitum* (pg. 43)

First antennal segment relatively short (SI < 30); eye separated from posterolateral clypeal margin by a distance less than or equal to width of antennal scape (as in Figs. 65-69) 7

7. (6) Dorsal petiolar node low and rounded (as in Fig. 53); ocelli small and emerging only slightly above posterior margin of head in full face view (Figs. 67-68); posterior margin of propodeum in lateral view straight to slightly convex 8

Dorsal petiolar node relatively tall and scale-like (as in Fig. 52); ocelli large and emerging strongly above posterior margin of head (Figs. 65-66); posterior margin of propodeum in lateral view straight to slightly concave 9

8. (7) Dorsum of petiolar node concolorous with mesosomal dorsum; widespread in Central America and the Caribbean to southern Brazil .... *L. iniquum* (pg. 50)

Dorsum of petiolar node pale white and much lighter in color than mesosomal dorsum; Southeastern Brazil .... *L. leucomelas* (pg. 81)

9. (7) Head sculpture densely punctate, surface opaque; mandible usually with apical tooth much smaller than subapical tooth; Atlantic forest regions of Brazil, Paraguay and Argentina .... *L. pulex* (pg. 105)

Head sculpture not densely punctate, surface shining through pubescence; mandible with apical tooth long and subapical tooth reduced or absent 10

10. (9) Size larger, MML > 0.85; Costa Rica to southeastern Brazil .... *L. neotropicum* (pg. 91)

Size smaller, MML < 0.80; Bolivia and Paraguay to eastern Brazil .... *L. cerradense* (pg. 37)

11. (1) Digitus with distal arm short, less than 1/3 length of proximal arm 12

Digitus with distal arm long, greater than 1/2 length of proximal arm 13

12. (11) Dorsal edge of volsella and proximal arm of digitus straight to slightly concave (Fig. 62); eye size smaller (EL < 0.38, OI < 54); legs shorter (FL < 1.05); northwestern South America to Costa Rica .... *L. piliferum* (pg. 101)

Dorsal edge of volsella and digitus strongly concave (Fig. 64); eye size larger (EL > 0.38, OI > 54); legs longer (FL > 1.05); Ecuador and Colombia .... *L. tsachila* (pg. 109)

13. (11) Distal arm of digitus much longer than height of volsella in lateral view (Fig. 60); eyes relatively small (EL < 0.30); Hispaniola .... *L. keiteli* (pg. 76)

Height of volsella in lateral view subequal to length of distal arm of digitus (Figs. 56, 58); eyes larger (EL > 0.30) 14

14. (13) Proximal arm of digitus broad at base and triangular in shape (Fig. 56); ventrodistal process of volsella absent or present as a small tooth or weak spine; legs longer (FL > 1.1, FI > 69); Costa Rica to Brazil .... *L. angulatum* (pg. 29)

Proximal arm of digitus narrow at base and filamentous (Fig. 58); ventrodistal process of volsella present as a well developed spine; legs shorter (FL < 1.1, FI < 69); Peru and Ecuador .... *L. fuscum* (pg. 51)

Species Synopses

Linepithema anathema Wild, **sp. nov.**

(worker mesosoma Fig. 33; worker head Fig. 34; distribution Fig. 106)

Species group: Humile

Holotype worker. BRAZIL. Minas Gerais: 2 km S Monte Verde, 22°54'S 46°03'W, 1900m, 26.viii.1996, under stone in shrubland, P.S. Ward acc. no. PSW13155 [MZSP].

Paratypes. Same data as holotype, 9 workers [ALWC, BMNH, CASC, LACM, MCZC, MHNG, UCDC, USNM].

Holotype worker measurements: HL 0.64, HW 0.51, MFC 0.14, SL 0.64, FL 0.54, LHT 0.62, PW 0.36, ES 1.94, SI 125, CI 80, CDI 28, OI 30.

Worker measurements: (n = 5) HL 0.62–0.68, HW 0.51–0.56, MFC 0.14–0.15, SL 0.63–0.66, FL 0.54–0.58, LHT 0.59–0.66, PW 0.36–0.40, ES 1.94–2.25, SI 119–126, CI 80–85, CDI 25–28, OI 30–36.

Worker diagnosis: Head narrow in full face view (CI 80–85); antennal scapes relatively long (SI 119–126); mesopleura, metapleura, and all gastric tergites with dense pubescence; gastric tergite 2 bearing suberect to erect setae; mesosoma bicolored with dorsum brown and mesopleura and metapleura light reddish brown.

Worker description: Head in full face view ovoid and relatively narrow (CI 80–85). Lateral margins broadly convex, grading smoothly into posterior margin. Posterior margin convex. Compound eyes large (OI 30–36), comprising 75–90 ommatidia. Antennal scapes long (SI 119–126), approximately as long as HL. In full face view, scapes in repose exceeding posterior margin of head by a length greater than length of first funicular segment. Frontal carinae moderately to narrowly separated (CDI 25–28). Maxillary palps of moderate length, approximately ½ HL, ultimate segment (segment six) longer than segment 2.

Pronotum and mesonotum forming a continuous convexity in lateral view, mesonotal dorsum slightly convex, not angular or strongly impressed. Metanotal groove only slightly impressed. Propodeum in lateral view with dorsal and posterior faces subequal in length, posterior face only slight broken at level of spiracle.

Petiolar scale sharp and inclined anteriorly, in lateral view falling short of the propodeal spiracle.

Cephalic dorsum (excluding clypeus) without standing setae. Pronotum without standing setae. Mesonotum without standing setae. Gastric tergite 1 (= abdominal tergite 3) bearing 0–2 very short subdecumbent setae, tergite 2 with 2–3 suberect to erect setae, tergite 3 with 3–4 suberect to erect setae. Venter of metasoma with scattered erect setae.

Integument shagreened and only lightly shining. Body and appendages including gula, entire mesopleura, metapleura, and abdominal tergites covered in dense pubescence.

Normally somewhat bicolored. Head, mesosomal dorsum, gaster, legs, and antennae dark brown. Mandibles, mesopleura, and metapleura a light reddish brown.

Queen and male unknown.

Distribution: Southeastern Brazil.

Biology: The type series was collected under a stone in shrubland at 1900 meters elevation. Little is known about this species.

Similar species: Workers of the commonly encountered cosmopolitan species *L. humile* usually lack standing hairs on gastric tergites 1–2 and tend to have a somewhat broader head (CI 84–93). *Linepithema oblongum*, found in the high Andes of Bolivia and northern Argentina, is a very similarly proportioned ant to *L. anathema* but normally has at least some members of each series with dilute pubescence on gastric tergites 2–4. *Linepithema anathema* differs from both *L. humile* and *L. oblongum* in having a more upright, less anteriorly inclined propodeum.

Discussion: The two collections, from Minas Gerais and Paraná, are both similar in appearance. Males are not known, but they will probably key out to *L. micans* or *L. gallardoi.*

Etymology: From Latin, meaning "accursed thing." The long scapes, large eyes, and relatively sparse pilosity of this ant complicate the diagnosis of the morphologically similar pest species *L. humile.*

Material examined: BRAZIL. Minas Gerais: 2 km S Monte Verde, 22°54'S 46°03'W, 1900m [ALWC, BMNH, CASC, LACM, MCZC, MHNG, MZSP, UCDC, USNM]. Paraná: Castro [MZSP].

Linepithema angulatum (Emery), **stat. nov.**

(worker mesosoma Fig. 7; worker head Fig. 8; male head Fig. 55; male volsella Fig. 56; distribution Fig. 100)

Iridomyrmex humilis subsp. *angulatus* Emery 1894a: 165-166 (W, Q). Lectotype worker, by present designation [MCSN, examined], 14 worker and 2 queen paralectotypes, Salinas sul Beni, Bolivia, Balzan [MCSN, MHNG, examined; MCSN lectotype series contains additionally a single misidentified *Paratrechina* worker].
Iridomyrmex humilis subsp. *angulata* Emery. Emery 1912: 26.
Iridomyrmex pordescens Wheeler 1942: 214 (W). Lectotype worker, by present designation [MCZC, examined] and 8 worker paralectotypes, San José, Costa Rica, 1.xii.1911, W. M. Wheeler [MCZC, examined; lectotype series contains additionally a single misidentified *Paratrechina* worker]. **syn. nov.**
Linepithema humile angulatum (Emery). Shattuck 1992a: 16. First combination in
    Linepithema.
Linepithema pordescens (Wheeler). Shattuck 1992a: 16. First combination in
    Linepithema.
Linepithema humile angulatum (Emery). Shattuck 1994: 123.
Linepithema pordescens (Wheeler). Shattuck 1994: 127.
Linepithema humile angulatum (Emery). Bolton 1995: 246.
Linepithema pordescens (Wheeler). Bolton 1995: 247.

Species group: Fuscum

Worker measurements: (n = 33) HL 0.59–0.75, HW 0.54–0.72, MFC 0.15–0.20, SL 0.52–0.67, FL 0.46–0.60, LHT 0.47–0.62, PW 0.33–0.47, ES 1.21–2.08, SI 90–104, CI 87–96, CDI 26–30, OI 20–28.

Worker diagnosis: Metanotal groove strongly impressed; dorsal face of propodeum straight to slightly concave; mesopleuron and metapleuron lacking pubescence and strongly shining; cephalic dorsum posterior of clypeus usually lacking standing setae (occasionally with a pair of subdecumbent setae in Central America); head width > 0.53; color testaceous to medium brown.

Worker description: Head in full face view somewhat longer than broad (CI 87–96), lateral margins convex, posterior margin slightly concave to slightly convex, often straight. Head normally reaches widest point posterior of compound eyes. Compound eyes of moderate size (OI 20–28), comprised of 45–70 ommatidia. Antennal scapes relatively short (SI 90–104), shorter than head length. In full face view, scapes in repose exceed posterior margin of head by a length less than or subequal to length of first funicular segment. Frontal carinae moderately spaced (CDI 26–30). Maxillary palps moderately short, approximately ½ HL or less, ultimate segment (segment 6) shorter than segment 2.

Pronotum and anterior mesonotum forming a continuous curve. Mesonotal dorsum relatively angular, anterior dorsal face often medially impressed or concave. Metanotal groove strongly impressed. Propodeum relatively high and globose, dorsal propodeal face usually straight to slightly concave in lateral view. Posterior propodeal face convex.

Petiolar scale relatively sharp and inclined anteriorly, in dorsal view broad, in lateral view falling short of propodeal spiracle.

Cephalic dorsum (excluding clypeus) usually lacking erect setae, rarely with a single pair of subdecumbent setae near vertex (Central America). Pronotum with 1–3 erect setae (mean = 2.1). Mesonotum without erect setae. Gastric tergite 1 (= abdominal tergite 3) bearing 0–2 erect setae (mean = 1.7) mesally, exclusive of a row of 4–5 subdecumbent setae along posterior margin of tergite, tergite 2 bearing 2–4 erect setae (mean = 3.0) exclusive of posterior row, tergite 3 bearing 2–6 erect setae (mean = 4.3) exclusive of posterior row. Venter of metasoma with scattered erect setae.

Sculpture on head and mesosomal dorsum shagreened and relatively opaque. Pubescence dense on head, mesosomal dorsum, anterior petiolar scale, and gastric tergites 1–4. Mesopleura and metapleural bulla without pubescence and strongly shining.

Color usually concolorous testaceous, occasionally medium brown.

Queen measurements: (n = 4) HL 0.90–0.97, HW 0.83–0.90, SL 0.75–0.85, FL 0.82–0.93, LHT 0.93–1.05, EL 0.31–0.34, MML 1.98–2.09, WL 5.98–6.08, CI 92–95, SI 90–95, OI 34–36, WI 29, FI 41–47.

Queen description: Relatively large species (MML > 1.9). Head longer than broad in full face view (CI 92–95), posterior margin slightly concave. Eyes of moderate size (OI 34–36). Ocelli of moderate size. Antennal scapes of moderate length (SI 90–95), in full face view scapes in repose surpassing posterior margin by a length less than length of first funicular segment.

Forewings moderately short relative to mesosomal length (WI 29). Forewings with Rs+M somewhat longer than M.f2. Legs moderate to short relative to mesosomal length (FI 41–47).

Dorsum of mesosoma and metasoma with scattered fine erect to subdecumbent setae, mesoscutum usually with more than 10 standing setae. Body color medium reddish-brown. Antennal scapes and femora concolorous with body, coxae and tibiae usually lighter in color.

Male measurements: (n = 4) HL 0.73–0.77, HW 0.67–0.71, SL 0.20–0.21, FL 1.13–1.25, LHT 1.09–1.25, EL 0.32–0.35, MML 0.76–0.95, WL 4.1–4.75, PH 0.28–0.32, CI 92–95, SI 27–30, OI 43–47, WI 26–28, FI 70–73.

Male diagnosis: Forewing with 2 submarginal cells; volsella with distal arm greater than 2/3 length of proximal arm; proximal arm of digitus broad at base and triangular in shape; eyes relatively large (EL > 0.30); legs relatively long for Fuscum-group species (FI > 70).

Male description: Head slightly longer than broad in full face view (CI 93–95). Eyes relatively large (OI 43–47), occupying much of anterolateral surface of head and separated from posterolateral clypeal margin by a length less than width of antennal scape. Ocelli large and in full frontal view set above adjoining posterolateral margins. Antennal scape of moderate length (SI 27–30), about 2/3 length of 3rd antennal segment. Anterior clypeal margin broadly convex medially. Mandibles large and worker-like, masticatory margin broad, much longer than inner margin, bearing 1–4 apical teeth followed by alternating series of teeth and denticles, similar to worker dentition. Inner margin and exterior lateral margin strongly diverging.

Mesosoma moderately developed, shorter in length than metasoma. Mesoscutum slightly enlarged, not projecting strongly forward or overhanging pronotum. Scutellum large, convex, nearly as tall as mesoscutum and projecting well above level of propodeum. Propodeum in lateral view low and not overhanging petiole, dorsal face rounding evenly into posterior face, posterior face straight to convex. Forewings long relative to mesosomal length (WI 26–28) and bearing two submarginal cells. Wing color clear to slightly smoky with darker brown veins and stigma. Legs long relative to mesosoma length (FI 70–73).

Petiolar node bearing a blunt, broadly-rounded scale, node height less than node length. Ventral profile of node straight to slightly convex and lacking a distinct process. Gaster elongate in dorsal view, about 3 times as long as broad. Gonostylus produced as a slender filament. Volsella with ventrodistal process absent or present as a small tooth. Cuspis absent. Digitus elongate, distal arm long, at least 2/3 length of proximal arm, and slightly longer than height of volsella in lateral view. Proximal arm wide at base, nearly as tall as adjoining volsella, triangular in shape, and narrowing to juncture with distal arm.

Dorsal surfaces of body with scattered erect setae, mesoscutum with more than 6 erect setae. Venter of gaster with scattered setae. Pubescence dense on body and appendages, becoming sparse only on medial propodeal dorsum.

Head, body and appendages medium brown in color.

Distribution: Costa Rica south to the Brazilian Pantanal.

Biology: Collection records of *L. angulatum* range across a broad array of habitats from sea level to over 2800 meters. Two collections are from leaf litter in montane forest, one under a stone in a montane forest edge, one under bark in second-growth forest, one in rotting wood, and two in high elevation urban parks in Colombia and Ecuador. Wheeler (1942: 214) reports this species (as "*pordescens*") nesting in a *Tillandsia* bromeliad in Costa Rica. This species has been observed recruiting to tuna baits placed 30cm underground in Peru (M. Frederickson, pers. com), to surface sausage baits in Venezuela, and to surface tuna baits in the Pantanal (Orr et al. 2003). One collection records *L. angulatum* tending pseudococcids on *Croton gossypifolia* in Colombia and another tending root-aphids in a nest in Ecuador. Male alates have been taken in June in Ecuador and in December in Colombia.

A series of studies by Orr and various coauthors (Orr and Seike 1998, Orr et al. 2001, Orr et al. 2003) documented the interactions between *L. angulatum* (as "*L. humile*" or "*L. piliferum*") and the phorid parasitoid *Pseudacteon lontrae* Mattos and Orr 2002 in the Pantanal. The presence of phorid flies significantly alters the behavior of this species in the field, leading to changes in the ecological dominance hierarchy in the ant community.

Linepithema angulatum has been intercepted several times with orchids in various ports of entry into the U.S., suggesting that it has the opportunity to spread further with human commerce.

Similar species: Linepithema fuscum workers are smaller (HW > 0.53), usually darker in color, with relatively longer scapes (SI 102–108) and a much less impressed metanotal groove (Fig. 13). Workers of *L. piliferum* have standing setae on the cephalic dorsum (also present in some Central American *L. angulatum*) and longer antennal scapes (SI 99–120, Fig. 77). *Linepithema tsachila* has standing setae on the cephalic dorsum (also present in some Central American *L. angulatum*) and a broader head with a deeply concave posterior margin (Fig. 20). *Linepithema pulex* from the

Atlantic forest region is usually smaller, with a more opaque head and sparse pubescence on gastric tergites 3–4.

Discussion: Some worker specimens from Central America, including Wheeler's types of *pordescens*, present particular difficulty for the diagnosis of this species as they occasionally have a small pair of subdecumbent setae on the cephalic dorsum. This setal character is easy to see in well-curated specimens and in South America its absence is the most reliable character for separating *L. angulatum* from the similar but more pilose *L. piliferum* and *L. tsachila*. Central American specimens are variable in size and in color, some specimens have large eyes with nearly 70 ommatidia (elsewhere < 60), but otherwise are similar in body proportion and in the diagnostic shape of the propodeum and metanotal groove. Pubescence varies within the species without much geographic structure from fine and appressed to longer and somewhat wooly. Specimens from parts of Colombia and Costa Rica are frequently darker in color than specimens from elsewhere.

The species boundaries of *L. angulatum* are still somewhat unclear. The broad sympatry of *L. angulatum* with the closely related species *L. piliferum*, *L. tsachila* and *L. fuscum* generally supports the present scheme of recognized species, but the allopatric variation among specimens placed in *L. angulatum* is more problematic. Material collected south of Colombia is relatively uniform in spite of the probable paraphyly of South American *L. angulatum* with respect to the morphologically distinct *L. cryptobioticum* (Wild, molecular data). However, variation among the relatively sparsely collected Central American populations, and between Central American and South American populations, suggests that further division of this group may become necessary. Male genital characters and DNA sequence data have proven useful in separating species in the *fuscum*-complex, but males of *L. angulatum* are only known from South America and freshly-collected Central American specimens were not available for molecular study. Given that most variation is allopatric, I feel it is preferable to treat these ants as a single species until more material becomes available.

Material examined: BOLIVIA. Cochabamba: 109 km E Cochabamba, 17°10'S 65°44' W, 1400m [WPMC]. El Beni: Salinas sul Beni [MCSN, MHNG]. BRAZIL. Amazonas: S. Gabriel [MZSP]. Mato Grosso do Sul: Corumbá, Serra do Urucum [MZSP]; Passo do Lontra [UCDC]. Mato Grosso: M'pio Poconé Transpantaneira km 115, Base de Pesquiza I.B.B.F. [MZSP]. COLOMBIA. Boyacá: Villa de Leiva, Instituto Humboldt [ALWC]. Caquetá: Puerto Rico [WPMC]. Cundinamarca: Bogotá - Villavicencio Hwy Km 79 [MCZC]; Bridge Quebrada Blanca [WPMC]; Cáqueza [WPMC]; Chipaque [WPMC]; D. E. Bogotá, 2600m [CASC]; Fosca (as "Fossa") [WPMC]; Guayabetal [WPMC]; Puente Quetame [WPMC]; Quebrada Chirihara km 81, Bogota to Villavicencio, 1400m [MCZC]; Quebrada Susamuko, 23 km NW

Villavicencio, 1000m [MCZC]; Santa Fe de Bogotá [WPMC]. Huila: 3 km E Rivera [WPMC]; 8 km S Neiva [WPMC]; 8 km W La Plata [WPMC]; Conc. J. Villamil (Gigante) [WPMC]; Gallego, 21 Km W La Plata [WPMC]; La Plata [WPMC]; La Vega [WPMC]. Meta: 14 km N Villavicencio [WPMC]; Border of Meta & Cundinamarca [WPMC]; El Castillo [WPMC]; S. Villavicencio, Caño El Buque, 480m [MCZC]. Putumayo: Villa Garza, 460m [LACM]. Valle de Cauca: Cali [UCDC]. Colombia (s.loc.), various Port-of-entry U.S. intercepts [LACM, USNM]. Colombia (s.loc.) [NHMW]. COSTA RICA. San José: San José [MZSP]. Heredia: Santo Domingo, 09°59'N 84°05'W, 1100m [JTLC]. Puntarenas: Parq. Nac. Corcovado, Sirena, 08°28'N 83°35'W, 1-100m [BMNH, JTLC]; Osa Penn, C. Helado, 17 km N Rincon, 08°46'N 83°25'W, 10m [WPMC]; Manuel Antonio N.P., 09°23'N 84°09'W, 10m [PSWC]. ECUADOR. Napo: 6 km SE Archidona, Monteverde Ecological Reserve, 00°56'S 77°45'W, 620m [ALWC]; Guagua Sumaco, 44km down Hollin-Loreto Rd [UCDC]; Misahualli [MCZC]. Pichincha: Quito, 00°11'S 78°30'W, 2810m [QCAZ]; Quito, Campus of the Catholic University (PUCE), 00°13'S 78°30'W, 2800m [ALWC, BMNH, CASC, IFML, LACM, MCZC, QCAZ, USNM]. Tungurahua: 2 km E Rio Negro, 01°25'S 78°11'W, 1200m [ALWC, MCZC, MHNG, MZSP, NHMB, QCAZ, UCDC]. Ecuador (s.loc.), Port-of-entry U.S. intercepts [USNM]. PANAMA. Chiriquí: Bocas del Toro, Cont. Div. [WPMC]; Boquete [USNM]; 20.4 km N San Felix, 950m [WPMC]. PERU. Huánuco: Tingo Maria & vic. [MCZC]; Tingo Maria, 1k E of town, 600m [BMNH]. Lima: La Molina, nr. Lima [MCZC]; Lima [USNM]. Loreto: Res. Nat. Allpahuayo-Mishana, 03°58'S 73°25'W [ALWC]. VENEZUELA. Aragua: Parque Nacional H. Pittier, Rancho Grande [WPMC]; Rancho Grande, 1100m [MCZC, UCDC]. Distrito Federal: Inst. Estud. Avan., Caracas [WPMC]. Trujillo: 15 km ESE Bonocó, 09°11'N 70°09'W, 1160m [PSWC].

Linepithema aztecoides Wild, **sp. nov.**

(worker body Fig. 43; worker head Fig. 44; distribution Fig. 105, live workers Fig. 110)

Species group: unassigned

Holotype worker. PARAGUAY. Canindeyú: Reserva Natural del Bosque Mbaracayú, Lagunita, 24°08'S 55°26'W, 220m, 13.xi.2002. A.L.Wild acc. no. AW1683. [1 worker, INBP].

Paratypes. Same collection data as holotype, A. L. Wild acc. nos. AW1683, AW1686 [31 workers, ALWC, BMNH, CASC, IFML, JTLC, LACM, MCZC, MHNG, MZSP, NHMB, NHMW, QCAZ, UCDC, USNM].

Holotype worker measurements: HL 0.66, HW 0.57, MFC 0.17, SL 0.60, FL 0.49, LHT 0.55, PW 0.37, ES 1.57, SI 105, CI 86, CDI 30, OI 24.

Worker measurements: (n = 16) HL 0.61–0.69, HW 0.52–0.60, MFC 0.14–0.18, SL 0.57–0.70, FL 0.46–0.61, LHT 0.52–0.64, PW 0.32–0.40, ES 1.35–1.93, SI 99–117, CI 85–91, CDI 25–31, OI 21–28.

Worker diagnosis: Head, mesosoma, and petiole dorso-ventrally flattened; posterolateral corners of propodeum formed by propodeal spiracles; color reddish brown to dark brown.

Worker description: Head in full face view chordate, widest at or posterior to compound eyes, and narrowed anteriorly. Posterior margin concave. Clypeus relatively larger than in other *Linepithema* species, projecting forward such that overall clypeal length, measured from level of posterior margin between antennal insertions to level of anterior projection, is about half total clypeal width. Anterior clypeal margin with a relatively narrow but concave medial excision. Compound eyes of moderate size (OI 21–28) and comprised of 65–80 ommatidia (mean = 70). Antennal scapes long (SL 0.57–0.70; SI 99–117), slightly shorter than head length (Paraguay and southeastern Brazil) to slightly longer than head length (Mato Grosso) and projecting well beyond posterior margin with head in full face view. Frontal carinae moderately to widely separated (CDI 25–31) and terminate near midpoint level of compound eye. In lateral view, head noticeably dorso-ventrally flattened, dorsal and ventral faces straight and parallel to each other. Maxillary palps relatively long, greater than ½ HL, ultimate segment (segment six) longer than segment 2.

Mesosoma dorso-ventrally compressed. Pronotum and mesonotum with dorsal surfaces nearly flat. In lateral view, mesonotal profile nearly straight, without mesal impression. Propodeum in lateral view short, dorsal face nearly straight, much longer than posterior face, and declining posteriorly. Posterolateral corners of propodeum formed by propodeal spiracles. Posterodorsal face of propodeum between spiracles broadly concave.

Petiole with a broad, low, anteriorly-inclined dorsal scale. Base of scale in lateral view longer than scale height. Anterior face of scale rounded in lateral view, posterior face straight. Ventral process of petiole evenly convex.

Gaster dorso-ventrally compressed and triangular on posterior view. In lateral view, gastric tergite 1 (= abdominal tergite 3) somewhat depressed, anterior face considerably shorter than dorsal face, such that less of gastric volume appears contained under 1st gastric tergite than in most other *Linepithema* species.

Cephalic dorsum often with a single pair of short, suberect setae near vertex, rarely with a single pair of short erect setae just posterior of antennal insertions. Pronotum bearing a pair of long erect setae. Gastric tergite 1 (= abdominal segment 3) with 2–6 setae along the posterior margin, tergite 2 with 6–10 setae along the posterior margin and 1–2 setae elsewhere, tergite 3 with a posterior row of setae and 1–4 setae elsewhere, tergite 4 with a pair of long setae. Venter of metasoma with scattered erect setae.

Pubescence dense on most of body and on all gastric tergites. Pubescence sparse on posterolateral face of pronotum, on mesopleuron, on metapleuron, and on posterior propodeal face. Surface of body shagreened and lightly shining.

Color medium reddish brown to dark brown over body, legs, and antennae. Mandibles reddish brown. Trochanters and tarsi lighter, to nearly white.

Queen and male unknown.

Distribution: Central and southeastern Brazil from Mato Grosso to São Paulo and west into Paraguay.

Biology: The type series was collected by the author in humid subtropical low forest along the edge of a small cerrado in eastern Paraguay. The other localities suggest that this rarely-encountered ant inhabits low, open forests such as cerrados and riparian gallery forests. *Linepithema aztecoides* recruit to baits; the type series was collected at a honey bait, and the Chapada do Guimarães collection was at a sardine bait. Male and queen castes are not known, and there are no records of nests. In the field these ants run about with their gasters held above their heads (Fig. 110), giving them the appearance of a small monomorphic *Azteca*.

Similar species: Given the distinct flattened habitus of *L. aztecoides*, this species is unlikely to be confused with any other *Linepithema*.

Discussion: *L. aztecoides* varies considerably in body proportions across its range. Specimens from Mato Grosso, Brazil consistently have longer appendages (FL > 0.53; LHT > 0.58; SL > 0.63; SI > 110) than specimens from Goiás, São Paulo and Paraguay (FL < 0.53; LHT < 0.59; SL < 0.64; SI < 113), although there is some overlap. The Mato Grosso specimens are also less setose and redder in color. Given the allopatry of these variations and the scarcity of *L. aztecoides* specimens in collections, it is preferable to treat the different forms as a single species.

Etymology: In the field this species appears similar to ants in the genus *Azteca*.

Material examined: BRAZIL. Goiás: Alto Paraiso, Encosta "Ferro do Engomar" [MZSP]. Mato Grosso: Chapada das Guimarães, Cachoeira Pedra Furada [MZSP]; Utiariti, Rio Papagaio [MZSP]. São Paulo: Agudos [MZSP]; Cajurú, Fazenda Santa Carlota [MZSP]. PARAGUAY. Canindeyú: Reserva Natural del Bosque Mbaracayú, Lagunita, 24°08'S 55°26'W [ALWC, BMNH, CASC, IFML, INBP, JTLC, LACM, MCZC, MHNG, MZSP, NHMB, NHMW, QCAZ, UCDC, USNM]. Cordillera: Caacupé, Camp. J. Norment [ALWC].

Linepithema cerradense Wild, **sp. nov.**

(worker mesosoma Fig. 27; worker head Fig. 28; male head Fig. 66; distribution Fig. 104)

Species group: Neotropicum

Holotype worker. PARAGUAY. Canindeyú: Reserva Natural del Bosque Mbaracayú, Aguara Ñu, 24°11'S 055°17'W, 240m, 16.xi.2002, campo cerrado, nest in soil. A.L.Wild acc. nos. AW1687. [1 worker, INBP].

Paratypes. Same collection data as holotype, A. L. Wild acc. nos. AW1687–1689 [21 workers, ALWC, BMNH, LACM, MCZC, MHNG, MZSP, UCDC, USNM].

Holotype worker measurements: HL 0.57, HW 0.52, MFC 0.14, SL 0.50, FL 0.42, LHT 0.46, PW 0.33, ES 1.45, SI 96, CI 90, CDI 27, OI 25.

Worker measurements: (n = 17) HL 0.50–0.59, HW 0.44–0.54, MFC 0.11–0.14, SL 0.44–0.53, FL 0.38–0.45, LHT 0.39–0.49, PW 0.29–0.36, ES 1.14–1.57, SI 96–104, CI 85–90, CDI 24–28, OI 19–28.

Worker diagnosis: A small species (HW 0.44–0.54); head narrow (CI 85–90); antennal scapes in repose exceeding posterior margin of head by less than length of first funicular segment; mesosoma in lateral view compact and somewhat flattened dorsally, propodeum only slightly depressed below level of promesonotum; color testaceous to medium brown, never dark brown or black.

Worker description: Head in full face view narrow (CI 85–90), lateral margins convex, posterior margin straight. Compound eyes of moderate size (OI < 19–28), comprised of 50–65 ommatidia. Antennal scapes relatively short (SI 96–104), shorter than head length. In full face view, scapes in repose exceeding posterior margin of head by a distance less than length of first funicular segment. Frontal carinae relatively narrowly spaced (CDI 24–28). Maxillary palps of moderate length, approximately ½ HL, ultimate segment (segment 6) as long or longer than segment 2.

Mesosoma in lateral view compact and somewhat dorsally flattened relative to most other *Linepithema*. Pronotum and mesonotum forming a single continuous convexity, dorsal promesonotal profile slightly convex, mesonotal dorsum without central saddle or indentation. Metanotal groove slightly impressed to not at all impressed, propodeum only slightly depressed below level of promesonotum. Dorsal propodeal face relatively straight and sloping slightly downward, slope often subparallel to slope of dorsal face of promesonotum.

Petiolar scale relatively sharp and inclined anteriorly, in lateral view falling short of propodeal spiracle.

Cephalic dorsum (excluding clypeus) lacking erect setae. Pronotum with 0–2 erect setae (mean = 1.2). Mesonotum without erect setae. Erect setae on gastric tergites 1–4 (= abdominal tergites 3–6) sparse, tergite 1 bearing 0–2 erect setae (mean = 0.3), tergite 2 bearing 0–4 erect setae (mean = 1.8), tergite 3 bearing 0–5 erect setae (mean = 2.9). Venter of metasoma with scattered erect setae.

Sculpture on head and mesosomal dorsum shagreened and dull to slightly shining. Pubescence dense on head, mesosomal dorsum, anterior petiolar scale, and gastric tergites 1–4. Mesopleura and metapleural bulla without pubescence and strongly shining.

Color testaceous to medium brown, gaster often darker than mesosoma.

Queen measurements: (n = 2) HL 0.72–0.78, HW 0.62–0.66, SL 0.58–0.65, FL 0.58–0.66, LHT 0.62–0.70, EL 0.22–0.27, MML 1.27–1.66, WL 4.80, CI 85–86, SI 94–99, OI 31–34, WI 29, FI 40–46.

Queen description: Relatively small species (MML < 1.7). Head narrow in full face view (CI 85–86), posterior margin slightly concave. Eyes moderately small (OI 31–34). Ocelli small. Antennal scapes of moderate length (SI 94–99), in full face view scapes in repose surpassing posterior margin by a length less than length of first funicular segment.

Forewings moderately short relative to mesosomal length (WI 29). Forewings with Rs+M more than twice as long as M.f2. Legs moderate to short relative to mesosomal length (FI 40–46).

Dorsum of mesosoma and metasoma with standing setae sparse, mesoscutum bearing 1–4. Body color testaceous to medium brown. Antennal scapes, legs, and mandibles lighter.

Male measurements: (n = 1) HL 0.41, HW 0.38, SL 0.11, FL 0.40, LHT 0.36, EL 0.20, MML 0.74, WL n/a, PH 0.15, CI 93, SI 26, OI 47, WI n/a, FI 54.

Male diagnosis: Forewing with 1 submarginal cell; size small (HL < 0.45, MML < 0.80); petiolar node with dorsal scale taller than long in lateral view; posterior face of propodeum slightly concave, not strongly overhanging petiole; sculpture of head not well developed and surface shining through pubescence; mandibles with apical tooth unusually elongate.

Male description: Head longer than broad in full face view (CI 93). Eyes of moderate size (OI 47), occupying much of anterolateral surface of head and separated from posterolateral clypeal margin by a length less than width of antennal scape. Ocelli of moderate size and in full frontal view set above adjoining posterolateral margins. Antennal scape short (SI 26), about 3/4 length of 3rd antennal segment. Anterior clypeal margin convex medially. Mandibles moderately sized and somewhat elongate, masticatory margin broad, longer than inner margin, apical tooth enlarged as a sharp spine and followed by a series of 8–18 denticles. Inner margin and exterior lateral margin parallel to slightly diverging.

Mesosoma moderately developed and subequal in length to metasoma. Mesoscutum not greatly enlarged, projecting slightly forward over pronotum. Propodeum in lateral view not strongly overhanging petiole, posterior face slightly concave. Forewings bearing one submarginal cell. Wings transparent, with pale whitish yellow wing veins and stigma. Legs of moderate length relative to mesosoma (FI 54).

Petiolar node bearing an erect scale, node height taller than node length. Venter of node bearing a convex downward-pointing process. Gaster ovoid in dorsal view, about 2 times as long as broad. Gonostylus produced as a pointed, triangular pilose lobe. Volsella with cuspis present, digitus short and downturned distally.

Dorsal surfaces of body with erect setae sparse, mesoscutum lacking standing setae. Venter of gaster with scattered setae. Pubescence dense on body and appendages, becoming sparse only on medial propodeal dorsum. Sculpture on head and mesoscutum not well developed, surface shining through pubescence.

Head, mesosoma and metasoma testaceous. Mandibles, antennae, and legs pale whitish-yellow to light brown, lighter than body.

Distribution: Bolivia, Paraguay, eastern Brazil.

Biology: Linepithema cerradense is frequently associated with cerrado habitat. Of seven records where explicit habitat information is recorded, four are from cerrado. Two additional records are from roadsides and one is from rainforest edge. Most collections unaccompanied by habitat data are from regions where cerrado is the predominant habitat type (e.g., Res. Biol. Águas Emendadas, Brazil). *L. cerradense* have been taken at baits in Goiás, Brazil. Alates have been collected in December flying to a light in Santa Cruz, Bolivia.

I observed two colonies in a Paraguayan cerrado in the Mbaracayú Reserve in November 2002. Nests were in located in open soil and appeared inconspicuous, with an entrance hole scarcely wider than an individual ant surrounded by small piles of excavated earth. Colonies maintain multiple nest entrances or multiple nests, as there were dilute trails of ants across the surface connecting several of the entrances within a 1 meter^2 area. Nests are apparently fairly deep, as excavations near the surface failed to yield any chambers or brood. The nesting biology appears similar to that of the closely-related *L. neotropicum*. Workers were observed carrying pieces of dead arthropods back to their nests, including carcasses of male and worker *Pheidole* ants.

Similar species: Workers of *L. neotropicum* have a more convex promesonotum, a more depressed propodeum (Fig. 29), a broader head, longer antennal scapes (Fig. 81), and are often darker in color. Workers of the smaller species in the Humile-group appear superficially similar, but these have dense pubescence on the metapleura (as in Fig. 5). Workers of *L. pulex*, a similarly small, light-colored ant from the Atlantic forest region have a much more deeply impressed metanotal suture (Fig. 45). Males of *L. neotropicum* are similar in structure to *L. cerradense* males but tend to be darker in color and larger in size (MML > 0.90, Fig. 93).

Discussion: This species exhibits little variation across its range. Some worker specimens from São Paulo are less pilose than specimens from elsewhere, often lacking pronotal setae, and color appears to vary slightly among sites without much geographic structure to the variation. This species is closely related to, and possibly evolved from, the variable and widespread *L. neotropicum* (Wild, unpublished molecular data). Perhaps not surprisingly, populations of *L. neotropicum* from northeastern Brazil share some features in common with *L. cerradense*, being lighter colored with somewhat shorter pronotal setae, but these ants are otherwise structurally and morphometrically closer to *L. neotropicum* than to *L. cerradense*.

Etymology: The name *cerradense* refers to the predominant habitat association of this species.

Material examined: BOLIVIA. Santa Cruz, 35 km SSE Flor de Oro, 13°50'S 60°52'W, 450m [MCZC, PSWC]; Buena Vista, 17°27'S 63°40'W, [UCDC]. BRAZIL. Bahia: Sta. Rita de Cássia [MZSP]. Distrito Federal: Res. Biol. Águas Emendadas [MZSP]. Goiás: Alvorada do Norte, Faz. Mattos [MZSP]; Anápolis [MZSP]; Campinaçu, Serra da Mesa, 13°52'S 48°23'W [MZSP]; Colinas do Sul, Serra da Mesa, 14°01'S 48°12'W [MZSP]; Parque Nacional das Emas, 20km W Chapadão do Ceu [LACM]. Mato Grosso do Sul: Campo Grande [MZSP]. São Paulo: Agudos [MZSP]; Assis Rd 333, Km 42 [MZSP]; Mata de Santa Genebra, 22°49'S 47°06'W [UCDC]. PARAGUAY. Canindeyú: Reserva Mbaracayú, Aguara Ñu, 24°11'S 55°17'W, 240m [ALWC, BMNH, IFML, LACM, MCZC, MHNG, MZSP, NHMB, UCDC, USNM]. Central: Capiatá, 25°21'S 57°25'W [ALWC]; Guarambaré, 25°29'S 57°27'W [ALWC, BMNH, CASC]; Luque, 25°16'S 57°34'W [ALWC, INBP, USNM]. Cordillera: Caacupé, Camp. J. Norment, 25°22'S 57°05'W [ALWC].

Linepithema cryptobioticum Wild, **sp. nov.**

(worker mesosoma Fig. 9; worker head Fig. 10; distribution Fig. 100)

Species group: Fuscum

Holotype worker. PARAGUAY. Boquerón: Enciso, 21°12'S 61°40'W, 3-6.xi.2001, M. Leponce & T. Delsinne # 4022-5/3, Dry chaco in sifted litter [1 worker, INBP].

Paratypes. PARAGUAY. Boquerón: Enciso, 21°12'S 61°40'W, M. Leponce & T. Delsinne #4022-5/3, #7597, #22836, Dry chaco in sifted litter [7 workers, ALWC, BMNH, LACM, MCZC, MHNG, MZSP, USNM].

Holotype worker measurements: HL 0.58, HW 0.54, MFC 0.17, SL 0.41, FL 0.41, LHT 0.43, PW 0.36, ES 0.78, SI 76, CI 93, CDI 31, OI 13.

Worker measurements: (n = 6) HL 0.56–0.58, HW 0.52–0.54, MFC 0.16–0.18, SL 0.40–0.42, FL 0.39–0.41, LHT 0.40–0.43, PW 0.35–0.36, ES 0.62–0.80, SI 76–81, CI 91–96, CDI 31–33, OI 11–14.

Worker diagnosis: Antennal scapes very short (SI 76–81); eyes small, with < 25 ommatidia; frontal carinae relatively widely spaced (CDI > 30); propodeum and petiolar scale unusually tall; mesopleura and metapleural bulla lacking pubescence and strongly shining.

Worker description: Head in full face view relatively broad (CI 91–96), lateral margins convex, posterior margin concave. Compound eyes small (ES < 0.8), comprised of 17–22 ommatidia. Antennal scapes short (SI 76–81), much shorter than head length. In full face view, scapes in repose not exceeding posterior margin of head. Frontal carinae relatively widely spaced (CDI 31–33). Maxillary palps short, less than ½ HL, ultimate segment (segment six) shorter than segment 2.

Mesosoma in lateral view with pronotum and mesonotum forming a single continuous convexity, mesonotum relatively angulate, without central saddle or indentation. Metanotal groove slightly impressed, propodeum high and relatively bulky, nearly reaching level of mesonotum. Dorsal face with a slight medial saddle-like impression, posterior margin convex dorsad of metapleural bulla.

Petiolar scale unusually tall, in lateral view reaching level of propodeal spiracle. Anterior and posterior faces of scale parallel.

Cephalic dorsum (excluding clypeus) lacking erect setae. Pronotum with 2–3 erect setae (mean = 2.2). Mesonotum usually without erect setae. Erect setae on gastric tergites 1–4 (= abdominal tergites 3–6) relatively abundant, tergite 1 bearing approximately 10 erect setae of varying length. Venter of metasoma with scattered erect setae.

Sculpture on head and mesosomal dorsum lightly shagreened and moderately shining. Pubescence dense on head, mesosomal dorsum, anterior petiolar scale, and gastric tergites 1–4. Mesopleura and metapleural bulla without pubescence and strongly shining.

Color pale yellow to reddish yellow.

Queen and male unknown.

Distribution: Known only from type locality in the Paraguayan chaco.

Biology: The few collections are all from sifted leaf litter at a single site in the Paraguayan dry chaco. Little is known about the biology of *L. cryptobioticum*. This species is probably subterranean, as small eyes, robust build, and yellow color are traits typical of hypogaeic ants.

Similar species: Linepithema cryptobioticum is a distinct ant unlikely to be confused with other species. *Linepithema flavescens*, known only from Haiti, has slightly larger eyes (> 25 ommatidia), a relatively narrow frontal carinal distance (CDI < 30), and lacks the enlarged petiolar scale. *Linepithema pulex* and *L. angulatum* are similar in color and in robust build, but these species have larger eyes (> 30 ommatidia, Fig. 80), longer antennal scapes (Fig. 77), and lack the enlarged petiolar scale.

Discussion. Males are not known, but they will probably key out with other Fuscum-group species.

Etymology: The name *cryptobioticum* refers to the inferred subterranean habits of this species.

Material examined: PARAGUAY. Boquerón: Enciso, 21°12'S 61°40'W [ALWC, BMNH, INBP, LACM, MCZC, MHNG, MZSP, USNM].

Linepithema dispertitum (Forel)

(worker mesosoma Figs. 4, 21; worker head Fig. 22; male head Figs. 70–72; male body Fig. 53; distribution Figs. 102, 107)

> *Iridomyrmex dispertitus* Forel 1885: 351-352 (W, M). Lectotype worker, by present designation [MHNG, examined], 13 worker and 2 male paralectotypes, Tecpán 7000', Chimaltenango, Guatemala, M. Stoll. [MHNG, NHMW, examined].
> *Iridomyrmex dispertitus* Forel. Forel 1908: 395-396 (M). Redescription of male.
> *Linepithema dispertitum* (Forel). Shattuck 1992a: 16. First combination in *Linepithema*.
> *Linepithema dispertitum* (Forel). Shattuck 1994: 122.
> *Linepithema dispertitum* (Forel). Bolton 1995: 247.

Species group: Iniquum

Worker measurements: (n = 44) HL 0.58–0.81, HW 0.49–0.75, MFC 0.15–0.22, SL 0.53–0.76, FL 0.43–0.65, LHT 0.47–0.71, PW 0.31–0.47, ES 1.26–2.18, SI 97–113, CI 83–94, CDI 26–33, OI 20–30.

Worker diagnosis: Cephalic dorsum with 6 or fewer erect setae (often lacking entirely); mesonotum with a slight medial impression; mesopleura and metapleura smooth and shining, with pubescence sparse to absent; medium brown to piceous in color; Central America and Hispaniola.

Worker description: Head in full face view oval to heart-shaped, varying from relatively narrow to relatively broad (CI 83–94), lateral margins convex, posterior margin slightly convex in some smaller specimens to distinctly concave in medium to larger specimens. Compound eyes of moderate size (OI 20–30), comprised of 47–80 (mean = 61) facets. Antennal scapes relatively long (SI 97–113), slightly shorter to slightly longer than head length. In full face view, scapes in repose exceeding posterior margin of head. Frontal carinae moderately to broadly spaced (CDI 26–33). Maxillary palps of moderate length, approximately ½ HL, ultimate segment (segment six) subequal in length to, or slightly longer than, segment 2.

Mesosoma in lateral view with pronotum and mesonotum forming a more or less single convexity, usually interrupted by a slight mesal impression on mesonotum. Mesothorax at impression wider than widest diameter of fore coxa. Metanotal groove slightly to deeply impressed. Propodeum variable in shape, from raised and evenly rounded to somewhat flattened dorsally. Metapleural bulla relatively swollen and protruding.

Petiolar scale inclined anteriorly, in lateral view falling short of propodeal spiracle.

Cephalic dorsum (excluding clypeus) bearing 0–6 (mean = 1.9) erect setae. Pronotum bearing 0–2 (mean = 0.4) erect setae. Mesonotum without erect setae. Gastric tergite 1 (= abdominal tergite 3) bearing 0–6 (mean = 0.6) erect setae, tergite 2 with 0–9 (mean = 2.8) erect setae, tergite 3 with 0–8 (mean = 3.9) erect setae. Venter of metasoma with scattered erect setae.

Surface of head and mesosomal dorsum smooth and relatively shining. Pubescence dense, short, and often subdecumbent to suberect on dorsum of head and mesosoma, sometimes presenting a distinct gray velvety sheen. Pubescence dense on gastric tergites 1–2, often dense to sometimes dilute on tergites 3–4. Mesopleura and metapleural bulla without pubescence or rarely with a few appressed hairs, surface strongly shining.

Concolorous medium brown to piceous.

Queen measurements: (n = 4) HL 0.80–0.94, HW 0.71–0.88, SL 0.67–0.83, FL 0.65–0.84, LHT 0.72–0.95, EL 0.23–0.26, MML 1.48–1.70, WL 4.20–4.36, CI 89–94, SI 90–95, OI 28, WI 28–30, FI 43–49.

Queen description: Moderately small species (MML < 1.7). Head longer than broad in full face view (CI 89–94), posterior margin slightly concave. Lateral margins often somewhat impressed anterior of eyes. Eyes small (OI 28). Ocelli small. Antennal scapes of moderate length (SI 90–95), in full face view scapes in repose surpassing posterior margin by a length greater than or equal to length of first funicular segment.

Forewings moderately short relative to mesosomal length (WI 28–30). Forewings with Rs+M more than twice as long as M.f2. Legs moderately long relative to mesosomal length (FI 43–49).

Pilosity variable on dorsum of mesosoma and metasoma, sparse to relatively abundant. Mesoscutum with 0–10 standing setae. Body color medium brown to piceous. Antennal scapes, legs, and mandibles concolorous with body.

Male measurements: (n = 6) HL 0.55 –0.62, HW 0.47–0.58, SL 0.17–0.53, EL 0.20 – 0.23, MML 0.93–1.01, WL 2.27–3.50, FL 0.57–0.65, LHT 0.57–0.65, CI 85–96, SI 32–86, OI 32–40, WI 24–35, FI 60–69.

Male diagnosis: Forewing with 1 submarginal cell; first antennal segment relatively long (SI > 31); propodeum with convex posterior face; eye separated from posterolateral clypeal margin by a distance greater than or equal to width of antennal scape; color medium to dark brown.

Male description: Extremely variable across range and difficult to characterize (see discussion). Head relatively narrow in full face view (CI 85–96), head shape ovoid or in some populations (Guatemala and El Salvador) approaching a chordate worker-like shape. Eyes small (OI 32–40) and separated from posterolateral clypeal margin by a length less than or equal to width of antennal scape. Ocelli small and in full frontal view emerging only slightly or not at all above adjoining posterolateral margins. Antennal scape long to very long (SI 32–86), approaching the worker condition in some populations (Guatemala and El Salvador), varying from 70% to 320% length of 3rd antennal segment. Anterior clypeal margin convex medially. Mandibles large and worker-like, masticatory margin broad, much longer than inner margin, bearing 1–2 apical teeth followed by alternating series of teeth and denticles, as in worker dentition. Inner margin and exterior lateral margin strongly diverging.

Mesosoma not well developed and subequal in length to metasoma. Mesoscutum not greatly enlarged, not projecting forward over pronotum. Propodeum in lateral view low and not overhanging petiole, posterior face slightly convex and rounding gradually into dorsal face. Forewings normally of moderate length (WI 24–26), rarely very long (WI 35, Hispaniola), and bearing one submarginal cell. Wings clear to slightly smoky with medium brown wing veins and stigma. Legs moderately long relative to mesosoma (FI 60–69).

Petiolar node bearing a sharp, worker-like forward inclined scale (Guatemala and El Salvador) or a low, rounded scale with node height shorter than node length (Hispaniola), or intermediate, with sharp, upright scale (Mexico). Venter of node bearing a convex lobe or downward-pointing process. Gaster ovoid in dorsal view, about 2 times as long as broad. Gonostylus produced as triangular pilose lobe. Volsella with cuspis present, digitus short and downturned distally.

Dorsal surfaces of body with erect setae sparse to absent, mesoscutum lacking standing setae. Venter of gaster with scattered setae. Pubescence dense on body and appendages, becoming sparse only on medial propodeal dorsum. Sculpture on head and mesoscutum not well developed, surface shining through pubescence.

Head, legs, antennal scapes, mesosoma and metasoma medium brown to dark brown. Tibiae, trochanters, and antennal apices lighter.

Distribution: Northern Mexico to Panama, with an isolated population above 2400 meters in the Cordillera Central of Hispaniola.

Biology: Linepithema dispertitum is primarily a montane forest ant but has been recorded across a broad range of habitats. This species has been collected from 130 to over 3,000 meters in elevation, with more than 90% of records across the range of *L. dispertitum* being from above 1,000 meters. Where habitat information has been recorded, 15 collections are from montane pine forest, five from montane rain forest, three from oak woodland, two from coffee plantations, and one each from tropical dry forest and rainforest.

Unlike the largely arboreal and closely related *L. iniquum*, *L. dispertitum* is more frequently found nesting in soil or rotting wood. 28 nest records are from under stones, nine are from rotting wood, eight are from orchids, two are from moss on a tree trunk, and one is from under bark. Several of the orchid records were overland port-of-entry intercepts of full colonies with dealate queens into the United States from southern Mexico, suggesting the potential of this species to spread beyond its native range. Alate males have been recorded in April and June in Veracruz, Mexico, in November in Guatemala, in January in El Salvador, and in July and September in the Dominican Republic.

Linepithema dispertitum is probably monogynous in some populations. Of ten nest excavations that I conducted in Guatemala and in the Dominican Republic, four nests contained a single dealate queen, and the remainder uncovered no queens. Molecular genetic data will be needed to confirm monogyny, and given the extensive interpopulation variation in male morphology it is possible that mating system and colony structure vary correspondingly.

I observed *Pseudacteon* sp. phorid flies attacking *L. dispertitum* at nest excavations in Baja Verapaz and Sololá, Guatemala. The flies appeared within a few minutes as I broke into the ant nests. Voucher specimens of the phorids were identified by Brian Brown and have been deposited at LACM.

In the Dominican Republic, *L. dispertitum* appears to be confined to a single population inhabiting pine forests above 2400 meters on Pico Duarte and neighboring mountains in the Cordillera Central. In these areas, *L. dispertitum* is abundant and apparently found to the exclusion of most other ant species.

Similar species: Workers of the closely related species *L. iniquum* usually have a much stronger mesonotal impression (Fig. 23) and always bear at least five and usually more erect setae on the cephalic dorsum posterior of the clypeus. Separating worker specimens without reference to geography can be difficult as both species are among the most variable in the genus over their full distribution. They overlap in nearly all measured morphometric characters, but in general *L. iniquum* specimens tend to have longer antennal scapes and a narrower pronotum in dorsal view than *L. dispertitum* (Fig. 82). Additionally, *L. iniquum* is normally arboreal while *L. dispertitum* tends to nest in soil or rotting wood. These ants co-occur only in southern Central America where they may be reliably differentiated by the sparse pubescence on gastric tergite 2 in *L. iniquum* versus moderate to dense pubescence on gastric tergite 2 in *L. dispertitum*, although these characters do not always hold elsewhere.

Discussion: Linepithema dispertitum is one of the more difficult species to diagnose owing to extensive variation among localities. This variation appears to be entirely allopatric, as there are no regions where more than one form is known to occur. Worker specimens from Costa Rica, the Dominican Republic, El Salvador, and Guatemala, including Forel's types of *L. dispertitum* from Tecpán, are relatively large-eyed (> 60 ommatidia), lack standing setae on the pronotum and often on the cephalic dorsum, show strong impression of the metanotal suture, and have dense pubescence on gastric tergites 1–4. Specimens from central Mexico have smaller eyes (OI < 26; < 60 ommatidia), a less-impressed metanotal suture, often with pubescence fading to sparse on abdominal tergites 3–4, and occasionally bear erect setae on the pronotum. Ants from the single northerly Chihuahuan collection have well developed pilosity, usually bearing more than 4 erect setae on the cephalic dorsum, a single pair of pronotal setae, and extensive pilosity on the gaster. The single collection from Baja California Sur is large (HW > 0.65), dark colored, and relatively pubescent, with a sparse pubescence extending onto the mesopleura and metapleural bulla.

Variation among the few known male specimens in this species is extraordinary and would suggest that this species be more finely divided if it were not for the continuous and allopatric nature of the variation. Guatemalan specimens from the type locality and from Sololá have an unusually worker-like head (Fig. 72) with short wings and long antennal scapes that easily surpass the posterior margin of the head in full face view. Specimens from Baja Verapaz, Guatemala and from La Libertad, El Salvador are similarly worker-like but with antennal scapes that are somewhat shorter. An unassociated male from Cuernavaca, Mexico has an even shorter first antennal scape that approaches the common condition of other *Linepithema* (Fig. 71). The Hispaniolan collections are similar to the Mexican specimen in scape length (Fig. 70) but have much longer wings (Fig. 98) and a lower, more rounded petiolar node (Fig. 53). Morphometric plots of scape length and scape index versus head length are given in Figures 94–96.

In spite of the extensive variation, it is unlikely that *L. dispertitum* is paraphyletic. Males share a number of apparently derived similarities, including the size and placement of the eyes, the long first antennal scape, the worker-like mandibles, and the relatively reduced structure of the mesosoma. Molecular data at several loci from populations in Baja California, Guatemala, and Hispaniola (Wild, unpublished) also support the monophyly of this species with respect to the close relative *L. iniquum*.

Material examined: COSTA RICA. San José: 4k E San Gerardo, 09°28'N 83°34'W, 2150m [PSWC, UCDC]. DOMINICAN REPUBLIC. Santiago: Loma Rucilla, 2500-3000m [MCZC]; Parque Nacional Armando Bermudez, Aguitas Frias, 19°02'N 70°56'W, 2700m [ALWC, BMNH, CASC, LACM, MCZC, MHNG, NHMW, UCDC, USNM]; Parque Nacional Armando Bermudez, La Comparticion, 19°02'N 70°58'W, 2500m [ALWC]; Parque Nacional Armando Bermudez, trail below Pico Duarte, 2820m [UCDC]; Summit La Pelona, 19°02'N 71°00'W, 3094m [UCDC]. EL SALVADOR. La Libertad: 6 km N of Tamanique, 1000m [MCZC]; Boquerón to Quezaltepeque, 1600m [MCZC]. GUATEMALA. Baja Verapaz: 4k WSW Purulhá, Llano Largo, 15°13'N 90° 16'W, 1800m [ALWC, BMNH, CASC, MCZC, MZSP, UCDC, USNM]. Chimaltenango: Tecpán, 2100m [MHNG, NHMW]. Sacatepéquez: Antigua [USNM]. Sololá: 3 km SSE San Andrés Semetabaj, 14°43'N 91°07'W, 2040m [ALWC, BMNH, CASC, IFML, LACM, MCZC, MHNG, MZSP, NHMW, UCDC, USNM]. Guatemala, Port-of-entry U.S. intercept [LACM]. Guatemala (s.loc.) [MCSN, NHMB, NHMW, USNM]. MEXICO. Baja California Sur: 3 km NE La Burrera, 23°31'N 110°02'W, 740m [UCDC]. Chihuahua: Hwy 66 at 44mi E Yecora, 28°26'N 108°30'W, 5250m [MCZC]. Jalisco: 14 km SSW Pto. Vallarta, 20°30'N 105°18'W, 130m [PSWC]; 6 km N El Tuito, 20°22'N 105°19'W, 730m [PSWC]. México: 2 km NE Tenancingo, 2200m [MCZC]. Michoacán: Port-of-entry U.S. intercept [USNM]. Morelos: Cuernavaca [MCZC, MHNG, USNM]; 5 mi E Cuernavaca [UCDC]; Derrame del Chichinautzin, 2250m [LACM]. Veracruz-Llave: Mirador [MCZC]; Coatepec [WPMC]; Sa. Teoviscocla, nr. Cuichapa, 1600m [MCZC]; Xalapa [WPMC]; 30 mi S Acayucan [UCDC]. Zacatecas: 7 mi. NW Sombrerete [LACM]. Mexico, various Port-of-entry U.S. intercepts [USNM]. PANAMA. Chiriquí: V. de Chiriquí, 1100m [BMNH].

Linepithema flavescens (Wheeler and Mann) **stat. nov.**

(worker mesosoma Fig. 11; worker head Fig. 12; distribution Fig. 107)

Iridomyrmex keiteli var. *flavescens* Wheeler and Mann 1914: 43 (W).
Lectotype worker, by present designation [MCZC, examined], and 5
worker paralectotypes, Cape Hatien, Haiti, W. M. Mann [MCZC,
examined].
Linepithema keiteli flavescens (Wheeler and Mann). Shattuck 1992a: 16. First
combination in *Linepithema*.
Linepithema keiteli flavescens (Wheeler and Mann). Shattuck 1994: 125.
Linepithema keiteli flavescens (Wheeler and Mann). Bolton 1995: 247.

Species group: Fuscum

Lectotype worker measurements: HL 0.71, HW 0.71, MFC 0.19, SL 0.58, FL 0.56,
LHT 0.55, PW 0.46, ES 1.28, SI 81, CI 101, CDI 27, OI 18.

Worker measurements: (n = 7) HL 0.64–0.72, HW 0.64–0.71, MFC 0.17–0.19, SL
0.54–0.58, FL 0.49–0.56, LHT 0.52–0.59, PW 0.43–0.46, ES 0.75–1.27, SI 81–85, CI
95–101, CDI 27, OI 11–18.

Worker diagnosis: Antennal scapes short (SI 81–85); eyes with < 40 ommatidia;
promesonotum strongly convex and relatively broad, mesonotum in lateral view
meeting dorsal face of propodeum at an angle of 90°–100°; color pale yellow.

Worker description: Head in full face view about as broad as long (CI 95–101), lateral
margins convex, posterior margin strongly concave. Compound eyes small (OI 11–
18), comprised of 28–40 ommatidia. Antennal scapes shorter than head length (SI 81–
85). In full face view, scapes in repose approximately reach posterior margin of head.
Frontal carinae moderately spaced (CDI 27). Maxillary palps of moderate length,
approximately ½ HL, ultimate segment (segment six) subequal in length to segment 2.

Mesosoma in lateral view with pronotum and mesonotum forming a single continuous
convexity, mesonotum strongly convex, without central saddle or indentation, and
meeting dorsal propodeal face at an angle of 90°–100°. Metanotal groove slightly
impressed, propodeum depressed well below level of mesonotum. Propodeum in
lateral view rounded, posterior margin convex dorsad of the metapleural bulla. In
dorsal view, mesosoma relatively broad (PW 0.43–0.46).

Petiolar scale inclined anteriorly, in lateral view falling short of propodeal spiracle.

Cephalic dorsum (excluding clypeus) lacking erect setae. Pronotum with 0–4 erect setae (mean = 1.3). Mesonotum without erect setae. Erect setae on gastric tergites 1–4 (= abdominal tergites 3–6) relatively sparse, tergite 1 bearing 10 erect setae of varying length. Venter of metasoma with scattered erect setae.

Sculpture on head and pronotal dorsum lightly reticulate-punctate and only slightly shining. Pubescence dense on head, propodeal dorsum, anterior petiolar scale, and gastric tergites 1–4. Mesonotal dorsum with moderate pubescence fading to sparse medially. Lateral face of pronotum, mesopleura and metapleural bulla without pubescence and strongly shining.

Color pale yellow.

Queen and male unknown.

Distribution: Hispaniola.

Biology: Little is known about the biology of this species. Mann collected the type series "under a stone on a dry hill-side at Cape Haitien" (Wheeler and Mann 1914: 43). The small eyes, short appendages, and pale coloration suggest a subterranean habit.

Similar species: Workers of *L. keiteli*, a commonly encountered Hispaniolan endemic, are closely similar in structure but have larger eyes (> 40 ommatidia), longer scapes (SI > 90, Fig. 79), more standing pilosity on the gaster, and are generally darker in color. Workers of *L. cryptobioticum*, a compact pale-colored ant from South America, have smaller eyes (< 25 ommatidia), a greater separation of the frontal carinae (CDI > 30), and a strongly elevated propodeum (Fig. 9).

Discussion: Linepithema flavescens is known from only two collections in northern and southern Haiti. Males have not been collected, but they will probably key out near *L. keiteli*. Worker specimens from Massif de La Hotte are considerably less pilose and have slightly smaller eyes (OI < 14) than specimens from the type locality at Cape Haitien. The Massif de La Hotte collection was made by Darlington in 1934. Unfortunately, this distinctive species has not been recorded subsequently and its conservation status is unknown.

Material examined: HAITI. Nord: Cape Haitien [MCZC]. Haiti (Dept unknown): La Hotte NE foothills [MCZC].

Linepithema fuscum Mayr

(worker mesosoma Fig. 13; worker head Fig. 14; male head Fig. 57; male volsella Fig. 58; distribution Fig. 103)

> *Linepithema fuscum* Mayr 1866: 497 (M). Lima, Peru. Lectotype male, by present designation [NHMW, examined], and 8 male paralectotypes [MCZC, NHMW, examined].
> *Iridomyrmex pilifer*. Crozier 1970: 113 (W). Karyotype; misidentification. [mat. ref. Machu Picchu, Peru, iii.1967, W. L. Brown, MCZC, examined].
> *Iridomyrmex* sp. nr. *pilifer*. Crozier 1970: 113 (W). Karyotype; misidentification. [mat. ref. Tingo Maria, Peru, 3.iii.1967, W. L. Brown , MCZC, examined].
> *Linepithema fuscum* Mayr. Shattuck 1992a: 16.
> *Linepithema fuscum* Mayr. Shattuck 1994: 122.
> *Linepithema fuscum* Mayr. Bolton 1995: 247.

Species group: Fuscum

Worker measurements: (n = 9) HL 0.53–0.58, HW 0.45–0.51, MFC 0.13–0.15, SL 0.48–0.54, FL 0.40–0.46, LHT 0.43–0.49, PW 0.30–0.32, ES 0.93–1.24, SI 102–108, CI 86–89, CDI 28–31, OI 17–23.

Worker diagnosis: Size small (HW < 0.52); cephalic dorsum posterior of compound eyes lacking standing hairs; mesopleura and metapleura lacking pubescence and strongly shining; eyes small (OI 17–23) with fewer than 50 ommatidia; body color medium brown to dark brown.

Worker description: Head in full face view relatively narrow (CI 86–89), lateral margins convex, posterior margin straight to slightly convex. Compound eyes small (OI 17–23), comprised of 35–50 facets. Antennal scapes of moderate length (SI 102–108), shorter than head length. In full face view, scapes in repose surpass posterior margin of head by an amount subequal to or slightly greater than the length of the first funicular segment. Frontal carinae moderately to widely separated (CDI 28–31). Maxillary palps of moderate length, approximately ½ HL, ultimate segment (segment six) shorter than, or subequal in length to, segment 2.

Mesosoma in lateral view compact, with pronotum and mesonotum forming a single continuous convexity, often interrupted by a slight mesal mesonotal impression. Mesonotum in lateral view angular. Metanotal groove not impressed or only slightly impressed. Propodeum relatively high, only slightly depressed below level of mesonotum, dorsal face relatively flat on top, meeting posterior face at an angle. Posterior propodeal face convex.

Petiolar scale sharp and inclined anteriorly, in lateral view falling short of the propodeal spiracle.

Cephalic dorsum (excluding clypeus) lacking erect setae or rarely with 1–2 subdecumbent setae near antennal insertions. Pronotum with 2–4 (mean = 2.3) erect to suberect setae. Mesonotum without erect setae. Gastric tergite 1 (= abdominal tergite 3) bearing 0–2 erect setae (mean = 1.6) exclusive of posterior row, tergite 2 with 2–5 (mean = 3.4) erect setae exclusive of posterior row, tergite 3 with 4–6 (mean = 4.4) erect setae. Venter of metasoma with scattered erect setae.

Surface of head and mesosomal dorsum shagreened and moderately shining. Pubescence dense and rather long on head and mesosomal dorsum. Mesopleura and metapleura lacking pubescence and strongly shining. Gastric tergites 1–4 with dense to moderate pubescence, surface moderately shining.

Body color medium brown to dark brown. Mandibles, antennae, trochanters and tarsi somewhat lighter.

Queen measurements: (n = 2) HL 0.86–0.87, HW 0.80–0.84, SL 0.76–0.78, FL 0.80–0.81, LHT 0.89–0.91, EL 0.32–0.35, MML 1.87–1.91, WL n/a, CI 92–98, SI 92–95, OI 37–40, WI n/a, FI 43.

Queen description: Moderately sized species (MML 1.87–1.91). Head longer than broad in full face view (CI 92–98), posterior margin slightly concave. Eyes large (OI 37–40). Ocelli moderately small. Antennal scapes of moderate length (SI 92–95), in full face view scapes in repose surpassing posterior margin by a length greater than or equal to length of first funicular segment.

Wings unknown (not present in examined material). Legs of moderate length relative to mesosomal length (FI 43).

Dorsum of mesosoma and metasoma with numerous standing setae. Mesoscutum with more than 10 standing setae. Body color medium brown. Antennal scapes, legs, and mandibles slightly lighter than body.

Male measurements: (n = 5) HL 0.68 –0.74, HW 0.63–0.70, SL 0.19–0.21, EL 0.32 –0.37, MML 1.48–1.67, WL 4.16–4.49, FL 1.01–1.07, LHT 1.01–1.07, CI 92–97, SI 27–30, OI 48–52, WI 26–28, FI 63–68.

Male diagnosis: Forewing with 2 submarginal cells; volsella with distal arm greater than 2/3 length of proximal arm; proximal arm of digitus narrow at base and filamentous in shape; eyes large (EL > 0.30); legs relatively short for Fuscum-group species (FI < 70).

Male description: Head slightly longer than broad in full face view (CI 92–97). Eyes relatively large (OI 47–52), occupying much of anterolateral surface of head and separated from posterolateral clypeal margin by a length less than width of antennal scape. Ocelli large and in full frontal view set above adjoining posterolateral margins. Antennal scape of moderate length (SI 27–30), about 70–80% length of 3rd antennal segment. Anterior clypeal margin broadly convex medially. Mandibles large and nearly worker-like, masticatory margin broad, much longer than inner margin, bearing 1–4 apical teeth followed by alternating series of teeth and denticles, similar to worker dentition. Inner margin and exterior lateral margin strongly diverging.

Mesosoma moderately developed, shorter in length than metasoma. Mesoscutum slightly enlarged, not projecting strongly forward or overhanging pronotum. Scutellum large, convex, nearly as tall as mesoscutum and projecting well above level of propodeum. Propodeum in lateral view not overhanging petiole, dorsal face rounding evenly into posterior face, posterior face straight to convex. Forewings long relative to mesosomal length (WI 26–28) and bearing two submarginal cells. Wing color clear to slightly smoky with darker brown veins and stigma. Legs long relative to mesosoma length (FI 62–68).

Petiolar node bearing a blunt, broadly-rounded scale, node height taller than node length. Ventral profile of node straight to slightly convex and lacking a distinct process. Gaster elongate in dorsal view, 2.5–3 times as long as broad. Gonostylus produced as a slender filament. Volsella with ventrodistal process present as a sharp spine. Cuspis absent. Digitus elongate, distal arm long, at least 2/3 length of proximal arm, and slightly longer than height of volsella in lateral view. Proximal arm narrow at base, less than ½ height of adjoining volsella, and filamentous in shape.

Dorsal surfaces of body with scattered erect setae, mesoscutum with more than 5 erect setae. Venter of gaster with scattered setae. Pubescence dense on body and appendages, becoming sparse only on medial propodeal dorsum.

Head, body and appendages medium brown in color.

Distribution: Ecuador, Peru.

Biology: Little is known about the biology of *L. fuscum*. This species has been collected from 200 to nearly 3000 meters. One lone male was collected in a "shrubby pasture", while a series from Madre de Dios was collected foraging inside a laboratory building. Crozier (1970) studied the karyotype of two populations of this species (called "*Iridomyrmex pilifer*" and "*I.* sp. nr. *pilifer*", vouchers at MCZC, examined), and found the haploid chromosome number of both populations to be N=9 and otherwise indistinguishable from each other.

Similar species: Linepithema angulatum workers also lack standing setae on the cephalic dorsum but are larger (HW > 0.53) and normally testaceous to light brown in color, with relatively shorter scapes (SI 90–104, Fig. 80) and a more deeply impressed mesonotal groove (Fig. 7). *Linepithema piliferum* and *L. tsachila* workers have standing setae on the cephalic dorsum near the vertex. Workers of *L. keiteli* from Hispaniola have a broader head (CI > 91) and a more shining integument. *Linepithema neotropicum* workers have larger eyes (> 50 ommatidia), a lower, more rounded propodeum, and at least some appressed pubescence on the metapleuron.

Discussion: Linepithema fuscum is the type species for the genus and was described by Mayr (1866) from males collected in Lima, Peru. Males and workers have never been definitively associated, so the placement of the small, dark-colored workers described here from various Peruvian localities should be regarded as tentative. This association is based on the following line of evidence. Five distinct male forms are known from the Fuscum species group, each diagnosable by their genitalia and each collected from more than one location. Four of these, *L. angulatum*, *L. keiteli*, *L. piliferum* and *L. tsachila*, have been associated with workers in at least one instance. Only *L. fuscum* itself remains without a worker association, and only one of the two remaining unassociated worker forms- the small dark form described above- is recorded from Peru. Both workers and males have been collected separately at a single location in Tingo Maria. Notably, the only other Fuscum-group species collected at Tingo Maria is *L. angulatum* whose male is associated in a William P. MacKay collection from Cundinamarca, Colombia. Given the variation within the latter species, however, it remains a possibility that the males of *L. fuscum* may actually pertain to the workers of *L angulatum*.

Workers show remarkably little variation among localities.

Material examined: ECUADOR. Napo: Guagua Sumaco, 45 km down Hollin-Loreto Rd [males, UCDC]. PERU. Amazonas: Chachapoyas, 2900m [male, MCZC]. Cusco: Machu Picchu [workers, MCZC]. Huánuco: Tingo Maria & vic. [males, queens, and workers, MCZC]; Monsoon Valley, Tingo Maria [worker, CASC]. Madre de Dios, 15 km NE Puerto Maldonado, 200m [workers, MCZC]. Lima: Lima [males, NHMW]. Peru (s.loc.) [males, NHMW].

Linepithema gallardoi (Brèthes)

(worker mesosoma Fig. 35; worker head Fig. 36; male head Fig. 75; distribution Fig. 106)

Dorymyrmex gallardoi Brèthes 1914: 95-96 (W). Type specimens not located, collection information given as Alta Gracia, Córdoba, Argentina, Gallardo.

Iridomyrmex humilis r. *platensis* var. *breviscapa* Forel 1914: 287 (W). Unavailable name (quadrinomen). Name made available by Santschi (1929: 306).

Iridomyrmex humilis subsp. *gallardoi* (Brèthes). Gallardo 1916 (W): 105-106. First combination in *Iridomyrmex*, first placement as infraspecific name in *humile*. Redescription of worker from primary type material.

Iridomyrmex humilis r. *platensis* var. *breviscapa* Forel. Santschi 1916: 390 (M). Unavailable name (quadrinomen). Name made available by Santschi (1929: 306).

Iridomyrmex impotens Santschi 1923: 67 (W). Lectotype worker, by present designation [NHMB, examined], Blumenau, Brazil, Reichensperger. **syn. nov.**

Iridomyrmex humilis st. *gallardoi* (Brèthes). Santschi 1929: 306 (W, Q).

Iridomyrmex humilis st. *breviscapa*. Santschi 1929: 306 (W). First available use of *breviscapa*. Misidentification of *L. micans* (Forel) [mat ref. 3w, NHMB, examined]. Lectotype worker, by present designation [MHNG, examined], and 1 paralectotype worker, Tucumán, Argentina, Shipton [MHNG, examined].

Linepithema humile breviscapum (Santschi). Shattuck 1992a: 16. First combination in *Linepithema*.

Linepithema humile gallardoi (Brèthes). Shattuck 1992a: 16. First combination in *Linepithema*.

Linepithema impotens (Santschi). Shattuck 1992a: 16. First combination in *Linepithema*.

Linepithema humile breviscapum (Santschi). Shattuck 1994: 123.

Linepithema humile gallardoi (Brèthes). Shattuck 1994: 124.

Linepithema impotens (Santschi). Shattuck 1994: 124.

Linepithema humile breviscapum (Santschi). Bolton 1995: 247.

Linepithema humile gallardoi (Brèthes). Bolton 1995: 247.

Linepithema impotens (Santschi). Bolton 1995: 247.

Species group: Humile

Worker measurements: (n = 20) HL 0.52–0.69, HW 0.43–0.65, MFC 0.13–0.17, SL 0.46–0.60, FL 0.39–0.54, LHT 0.40–0.56, ES 1.31–2.38, PW 0.29–0.44, CI 83–95, SI 89-107, CDI 24–29, OI 24–34.

Worker diagnosis: Mesonotum relatively robust, mesonotal dorsum usually angular with distinct dorsal and posterior faces; pubescence moderate to sparse on anterior mesopleuron; propodeum relatively high and inclined anteriorly, posterior propodeal face in profile broken at level of propodeal spiracle; antennal scapes short, in repose exceeding posterior margin of head by a length less than or subequal to that of first funicular segment.

Worker description: Head in full face view somewhat longer than broad (CI 83–95), lateral margins convex, posterior margin concave. Head normally reaches widest point at or posterior to level of compound eyes. Compound eyes of moderate size (OI 24–34), comprised of 50–85 ommatidia. Antennal scapes relatively short (SI 89–107, usually less than 100), shorter than head length. In full face view, scapes in repose exceed posterior margin of head by a length less than or subequal to that of first funicular segment. Frontal carinae narrowly to moderately spaced (CDI 24–29). Maxillary palps relatively short, approximately ½ HL or less, ultimate segment (segment 6) subequal in length to segment 2.

Pronotum and anterior mesonotum forming a continuous curve. Mesonotum relatively bulky, mesonotal dorsum angular and usually lacking a mesal impression, declivitous face rising sharply to meet dorsal face. Metanotal groove slightly to moderately impressed. Propodeum relatively high and inclined anteriorly, posterior propodeal face in profile broken at level of propodeal spiracle.

Petiolar scale relatively sharp and inclined anteriorly, in lateral view falling short of propodeal spiracle.

Cephalic dorsum (excluding clypeus) normally lacking erect setae, very rarely with a single pair of subdecumbent setae near vertex. Pronotum with 1–3 erect setae (mean = 2.1). Mesonotum without erect setae. Gastric tergite 1 (= abdominal tergite 3) bearing 2–8 erect to subdecumbent setae (mean = 4.4) including posterior row, tergite 2 bearing 2–8 erect setae (mean = 4.9), tergite 3 bearing 2–6 erect setae (mean = 3.5). Venter of metasoma with scattered erect setae.

Sculpture on head and mesosomal dorsum lightly shagreened and slightly to moderately shining. Pubescence dense on head, mesosomal dorsum, and gastric tergite 1. Pubescence variable from dense to sparse on gastric tergites 2–4, surface usually smooth and somewhat shining through pubescence. Mesopleuron with pubescence fading to moderate or sparse anteriorly, surface smooth and usually shining. Metapleuron with moderate pubescence.

Color variable, from light reddish brown to piceous, most commonly dark, mesosoma sometimes lighter than head and gaster.

Queen measurements: (n = 5) HL 0.78–0.81, HW 0.75–0.78, SL 0.64–0.67, FL 0.65–0.69, LHT 0.69–0.74, EL 0.24–0.28, MML 1.45–1.62, WL 4.45–4.64, CI 94–97, SI 85–89, OI 31–35, WI 30–32, FI 42–47.

Queen description: Relatively small species (MML < 1.7). Head slightly longer than broad in full face view (CI 94–97), posterior margin concave. Eyes of moderate size (OI 31–35). Ocelli of moderate size. Antennal scapes relatively short (SI 85–89), in full face view scapes in repose surpassing posterior margin by a length less than length of first funicular segment.

Forewings moderately short relative to mesosomal length (WI 30–32). Forewings with Rs+M subequal in length to M.f2. Legs of moderate length relative to mesosomal length (FI 42–47).

Dorsum of mesosoma and metasoma with numerous standing setae. Mesoscutum bearing 2–10 standing setae. Body color medium brown to piceous. Antennal scapes, legs, and mandibles concolorous with body, trochanters usually lighter.

Male measurements: (n = 6) HL 0.50–0.53, HW 0.47–0.51, SL 0.11–0.14, FL 0.47–0.56, LHT 0.41–0.50, EL 0.19–0.20, MML 0.95–1.08, WL 2.21–2.72, PH 0.18–0.23, CI 93–98, SI 23–27, OI 45–49, WI 23–25, FI 49–52.

Male diagnosis: Forewing with 1 submarginal cell; propodeum with strongly concave posterior face, overhanging petiole; mesosoma not greatly swollen (MML < 1.1); appendages not elongate (FI < 56); pubescence on head usually moderate to sparse in area posterior to compound eye.

Male description: Head slightly longer than broad to about as long as broad in full face view (CI 93–98). Eyes of moderate size (OI 45–49), occupying much of anterolateral surface of head anterior of midline and separated from posterolateral clypeal margin by a length less than width of antennal scape. Ocelli small and in full frontal view set above adjoining posterolateral margins. Antennal scape moderately long (SI 23–27), about 2/3 length of 3rd antennal segment. Anterior margin of median clypeal lobe broadly convex. Mandibles variable, small to moderate in size, usually bearing a single apical tooth and 8–13 denticles along masticatory margin. Masticatory margin relatively narrow to relatively broad, longer than or subequal in length to inner margin. Inner margin and exterior lateral margin converging, parallel, or diverging.

Mesosoma moderately developed, slightly larger or subequal in bulk to metasoma. Mesoscutum enlarged, projecting forward in a convexity overhanging pronotum. Scutellum large, convex, nearly as tall as mesoscutum and projecting above level of propodeum. Propodeum well developed and overhanging petiolar node, posterior propodeal face strongly concave. Forewings of moderate length relative to mesosomal length (WI 23–25) and bearing a single submarginal cell. Wing color clear to slightly smoky with light to dark brown veins and stigma. Legs moderately short relative to mesosoma length (FI 49–52).

Petiolar scale sharp and taller than node length. Ventral process well developed and pointing posteriorly. Gaster oval in dorsal view, nearly twice as long as broad. Gonostylus produced as bluntly rounded pilose lobes. Volsella with cuspis present, digitus short and downturned distally.

Dorsal surfaces of body largely devoid of erect setae, mesoscutum lacking standing setae, posterior abdominal tergites with a few fine, short setae. Venter of gaster with scattered setae. Pubescence dense on body and appendages, becoming sparse only on medial propodeal dorsum, lateral faces of pronotum, and often on the head posterior to the compound eyes.

Head, mesosoma and metasoma light to dark brown. Legs, mandibles, and antennae lighter in color.

Distribution: Venezuela and Colombia south to Argentina and southeastern Brazil.

Biology: Linepithema gallardoi is broadly distributed in South America from sea level to nearly 3000 meters in elevation and is most commonly recorded from open or low forest habitats. Nine records are from montane scrub grassland, three from grassland or pasture, two from urban lawns, one from pampas scrub, one from humid chaco, one from cleared cloud forest, one from subtropical forest edge, one from subtropical scrub forest, and one from a coffee plantation. This species has been collected in subterranean sausage traps and pitfall traps in Venezuela and at sardine baits in Brazil. Orr & Seike (1998) report this species (as "*L. humile*", voucher specimens at UCDC examined) being attacked by *Pseudacteon* phorid flies near Monte Verde, Minas Gerais, Brazil.

Eight nest records of this species are from under stones, two from open soil, and one from leaf litter. I excavated several colonies near the type locality at La Falda, Córdoba, Argentina. Most were polydomous with a diffuse series of small brood chambers connected through subterranean tunnels near the soil surface or under stones. Single dealate queens were seen in two of eight excavations, with no queens seen in the remaining six excavations, indicating that *L. gallardoi* may be monogynous, although molecular genetic work will be needed for confirmation.

Alates have been recorded in nests of *L. gallardoi* nests during October and January in Argentina, in November in Brazil, and in April in Colombia. Queens of this species are flighted, and Queens and males have been collected flying to lights in Paraguay in November and December.

Similar species: Workers of other species in the Humile-group always have more densely pubescent mesopleura and often longer antennal scapes (Fig. 85) and legs (Figs. 83–84). Workers of *L. neotropicum* have a lower, more rounded propodeum (Fig. 29) and longer maxillary palps.

Discussion: The type specimen of *L. gallardoi*, ostensibly at MACN, was not seen during this study and could not be located. The name was originally combined in *Dorymyrmex* by Brèthes (1914), but after viewing specimens Gallardo (1916) transferred *gallardoi* to *Iridomyrmex*. Unfortunately, neither the original description nor the figures and redescription by Gallardo (1916) are sufficient to confirm even the generic association of *gallardoi*. Gallardo's hastily prepared line drawing of the head seems to suggest that this ant may be as Brèthes suggested a *Dorymyrmex*, but the lateral view appears to show a *Linepithema*. The name *gallardoi*, at any rate, has been consistently applied in museum collections to the small, shiny, dark-colored *Linepithema* that is abundant at the type locality and in the Sierras del Córdoba of Argentina. Given that the brief description is not inconsistent with the common *Linepithema* from the region, I find it preferable to retain *gallardoi* in the current use. In the event that *gallardoi* proves inapplicable to the species described here, the name *breviscapa* Santschi 1929 remains available for this ant. I have designated lectotypes for *breviscapa* from Forel's original series for the unfortunate reason that Santschi made the name available based on a misidentification of *L. micans*.

Although all populations share the characteristic robust mesosoma and the shiny, sparsely pubescent anteroventral mesopleura, *Linepithema gallardoi* shows significant variation among populations in size, color, and pubescence. The extent of mesopleural pubescence can vary from nearly absent over the entire sclerite to moderately dense across the posterior 2/3 and fading to sparse anteriorly. While local populations are often consistently homogenous in their mesopleural pubescence, there is little broad scale geographic structure to the variation. Worker specimens from central Argentina, including the type locality of *gallardoi*, often have sparse gastric pubescence, although the variation in pubescence is continuous and a few Paraguayan and Brazilian specimens are also sparsely pubescent. Santschi's *impotens* type from Santa

Catarina is very sparsely pubescent on both the mesosoma and metasoma. Some of the more northerly montane populations, including those from northern Colombia and some from around Cusco, Peru, are relatively large and medium reddish-brown in color.

Material examined: ARGENTINA. Córdoba: 22 km WSW Alta Garcia, 31°44'S 64°39'W, 1200m [PSWC]; Alta Gracia [MZSP, NHMB]; Alta Gracia, La Granja, Sierras del Córdoba [MACN]; c. 20 km W Devoto, 31°24'S 62°34'W, 100m [ALWC, UCDC]; La Falda, 31°05'S 64°28'W, 1100m [ALWC, CASC, IFML, LACM, MHNG, MZSP]; La Falda, 31°05'S 64°27'W, 1250m [ALWC]; La Falda, 31°06'S 64°28'W, 1320m [ALWC, BMNH, MCZC, UCDC, USNM]; La Granja [NHMB]; La Paz, Dept. San Javier [MACN, NHMB]; Córdoba (s.loc.) [MACN]. Formosa: Pilcomayo, 25°04'S 58°05'W [UCDC]; Pilcomayo, 25°02'S 58°06'W [UCDC]. Tucumán: Tucumán [MHNG]. BOLIVIA. La Paz: Sector Pintada, 20-35 km NNW Apolo, 1650m [UCDC]. BRAZIL. Distrito Federal: Res. Biol. Águas Emendadas [MZSP]. Goiás: 7 km NW Alto Paraiso, Morro das Cobras [MZSP]; Anápolis [MZSP]. Minas Gerais: Monte Verde, 22°52'S 46°03'W [UCDC]; Sabará [USNM]; Serra Caraça, 1380m [MZSP]. Mato Grosso: Utiariti, Rio Papagaio [MZSP]. Paraná: 4 km E Rio Negro, Mun. Curitiba [MCZC]; Col. Esperança, Arapongas, 900m [MZSP]; Maringá, Poço das Antas [MZSP]; Maringá, Pousada Casarão [MZSP]; Rio Negro [NHMB]. Rio de Janeiro: Ilha da Marambaia [MZSP]; Itatiaia [MZSP]; Petrópolis [MZSP]. Rio Grande do Sul: S. Augusto, 500m [MZSP]. Santa Catarina: Blumenau [NHMB]; Canoinhas [MZSP]; N. Teutônia, 400m [MZSP]. São Paulo: Anhembi, Faz. B. Rico [MZSP]; Araçoiaba [MZSP]; Barueri [MZSP]; Campo Limpo [MZSP]; Ibiúna [MZSP]; Itanhaém [MZSP]; S. Cantareira [MZSP]; Teod. Sampaio [MZSP]. COLOMBIA. Boyacá: Sogamoso, 2500m [ALWC]. Cundinamarca: Bogotá [UCDC, WPMC]; Cáqueza [WPMC]. Magdalena: San Lorenzo, N. Sierra Nevada de Sta. Marta, 11°06'N 74°03'W, 2200m [PSWC, UCDC]. Risaralda: Apia, La Estrella, 1470m [ALWC]. Valle: Alcala [ALWC]; Bosque de Yotoco, 1575m [WPMC]. PARAGUAY. Canindeyu: Mbaracayú, 24°08'S 55°26'W, 220m [ALWC, BMNH, CASC, IFML, LACM, MCZC, MHNG, MZSP, UCDC, USNM]; Col. "11 de Setiembre", 25°02'S 55°34'W, 220m [ALWC]. Central: San Lorenzo, 25°21'S 57°31'W [ALWC, BMNH, MCZC, MZSP]. Guairá: Col. Independencia, ANTELCO, 25°43'S 56°15'W [ALWC]. Itapúa: San Miguel Portrero, c/ Villa Yacyreta, 27°02'S 56°12'W [ALWC, INBP]. Pte. Hayes: Ruta Trans-Chaco, 24°08'S 58°19'W, 120m [ALWC]. PERU. Amazonas: Chachapoyas, 2900m [MCZC]. Cusco: Machu Picchu ruins, 13°09'S 72°31'W, 2700m [ALWC, MCZC]. Huánuco: Huánuco airport [MCZC]. VENEZUELA. Distrito Federal: Inst. Estud. Avan., Caracas [WPMC].

Linepithema humile (Mayr)

(worker body Fig. 31; worker head Fig. 32; male head Fig. 74; male body Fig. 54; male wing Fig. 48; male volsella Fig. 50; distribution Figs. 101, 108).

Hypoclinea humilis Mayr 1868: 164 (W). Holotype worker, Buenos Aires, Argentina, Stroebel, 1866 [NHMW, examined].

Iridomyrmex humilis (Mayr). Emery 1888: 386-388. First combination in *Iridomyrmex*.

Iridomyrmex humilis (Mayr). Wheeler 1913a: 27-29 (M, Q, W). Male and queen description, worker redescription.

Iridomyrmex humilis (Mayr). Newell and Barber 1913: 38-39 (egg), 40-41 (larva), 42-45 (W, M, Q pupae).

Iridomyrmex humilis var. *arrogans* Chopard 1921: 241-245 (W). Lectotype worker, by present designation [NHMB, examined], Cannes, Provence-Alpes-Cote d'Azur, France, Chopard [NHMB, examined]. Junior synonym of *I. humilis* by Bernard 1967: 251. Restored to subspecies of *L. humile* (Mayr) by Shattuck 1992a: 16. Returned to synonymy by Wild 2004: 1207.

Iridomyrmex riograndensis Borgmeier 1928: 64 (W). Lectotype worker, by present designation [MZSP, examined] and 7 worker paralectotypes, Rio Grande do Sul (s.loc.), Brazil, 19.i.1918 [MZSP, examined]. Junior synonym of *L. humile* (Mayr) by Wild 2004: 1207.

Iridomyrmex humilis (Mayr). Wheeler and Wheeler 1951: 186-189. Summary of larval biology.

Iridomyrmex humilis (Mayr). Crozier 1969: 250. Karyotype.

Linepithema humile (Mayr). Shattuck 1992a: 16. First combination in *Linepithema*.

Linepithema humile arrogans (Chopard). Shattuck 1992a: 16. First combination in *Linepithema*.

Linepithema riograndensis (Borgmeier). Shattuck 1992a: 16. First combination in *Linepithema*.

Linepithema humile (Mayr). Shattuck 1994: 123.

Linepithema humile arrogans (Chopard). Shattuck 1994: 123.

Linepithema riograndensis (Borgmeier). Shattuck 1994: 127.

Linepithema humile (Mayr). Bolton 1995: 247.

Linepithema humile arrogans (Chopard). Bolton 1995: 246.

Linepithema riograndensis (Borgmeier). Bolton 1995: 247.

Linepithema humile (Mayr). Wild 2004: 1204-1215 (M, Q, W). Distribution and taxonomic redescription.

Species group: Humile

Holotype worker measurements: HL 0.74, HW 0.66, MFC 0.16, SL 0.76, FL 0.65, LHT 0.68, PW 0.45, ES 2.93, SI 115, CI 89, CDI 24, OI 40.

Worker measurements: (n = 81) HL 0.62–0.78, HW 0.53–0.72, MFC 0.14–0.18, SL 0.62–0.80, FL 0.52–0.68, LHT 0.57–0.76, PW 0.35–0.47, ES 1.98–3.82, SI 108–126, CI 84–93, CDI 23–28, OI 32–49.

Worker diagnosis: Eyes large (OI > 30); antennal scapes long (SI > 105); pronotum and first two gastric tergites lacking erect setae; mesopleura and metapleura densely pubescent.

Worker description: Head in full face view longer than broad (CI 84–93), narrowed anteriorly and reaching its widest point just posterior to compound eyes. Lateral margins broadly convex, grading smoothly into posterior margin. Posterior margin straight in smaller workers to weakly concave in larger workers. Compound eyes large (OI 32–49), comprising 82–110 ommatidia (normally around 100). Antennal scapes long (SI 108–126), as long or slightly longer than HL and easily surpassing posterior margin of the head in full face view. Frontal carinae narrowly to moderately spaced (CDI 23–28). Maxillary palps relatively short, shorter than ½ HL, ultimate segment (segment six) noticeably shorter than segment 2.

Pronotum and mesonotum forming a continuous convexity in lateral view, mesonotal dorsum nearly straight, not angular or strongly impressed, although sometimes with a slight impression in anterior portion. Metanotal groove moderately impressed. Propodeum in lateral view inclined anteriad. In lateral view, dorsal propodeal face meeting declivity in a distinct though obtuse angle, from which the declivity descends in a straight line to the level of the propodeal spiracle.

Petiolar scale sharp and inclined anteriorly, in lateral view falling short of the propodeal spiracle.

Dorsum of head (excluding clypeus), mesosoma, petiole, and gastric tergites 1–2 (= abdominal tergites 3–4) devoid of erect setae (very rarely with a pair of small setae on gastric tergite 2). Gastric tergites 3–4 each bearing a pair of long, erect setae. Venter of metasoma with scattered erect setae.

Integument shagreened and lightly shining. Body and appendages including gula, entire mesopleura, metapleura, and abdominal tergites covered in dense pubescence.

Body and appendages concolorous, most commonly a medium reddish or yellowish brown but ranging in some populations from testaceous to dark brown, never yellow or piceous.

Queen measurements: (n = 13) HL 0.83–0.92, HW 0.83–0.93, SL 0.81–0.89, FL 0.78–0.90, LHT 0.88–0.97, EL 0.31–0.36, MML 1.67–2.09, WL 4.42–4.51, CI 93–101, SI 96–102, OI 36–39, WI 24–27, FI 40–48.

Queen description: Moderately large species (MML 1.67–2.09). Head slightly longer than broad to as broad as long in full face view (CI 93–101), posterior margin slightly concave to slightly convex. Eyes of moderate size (OI 36–39). Ocelli small. Antennal scapes relatively long (SI 96–102), in full face view scapes in repose surpassing posterior margin by a length greater than length of first funicular segment.

Forewings short relative to mesosomal length (WI 24–27). Forewings with Rs+M at least three times longer than M.f2. Legs of moderate length relative to mesosomal length (FI 40–48).

Dorsum of mesosoma and metasoma with scattered standing setae. Mesoscutum bearing 2–11 standing setae. Body color medium reddish brown. Antennal scapes, legs, and mandibles concolorous with body.

Male measurements: (n = 12) HL 0.56–0.70, HW 0.56–0.74, SL 0.13–0.16, FL 0.60–0.77, LHT 0.51–0.66, EL 0.31–0.34, MML 1.40–1.96, WL 2.55–3.26, PH 0.25–0.34, CI 99–106, SI 22–27, OI 51–55, WI 17–20, FI 37–45.

Male diagnosis: Forewing with single submarginal cell; mesosoma robust (MML > 1.3), mesoscutum greatly enlarged and overhanging pronotum; wings short relative to mesosomal length (WI < 21).

Male description: Head about as broad as long in full face view (CI 99–106). Eyes large (OI 51–55), occupying much of anterolateral surface of head and separated from posterolateral clypeal margin by a length less than width of antennal scape. Ocelli large and in full frontal view set above adjoining posterolateral margins. Antennal scape of moderate length (SI 22–27), about 2/3 length of 3rd antennal segment. Anterior clypeal margin straight to broadly convex. Mandibles small, bearing a single apical tooth and 4-8 denticles along masticatory margin and rounding into inner margin. Masticatory margin relatively short, subequal in length to inner margin. Inner margin roughly parallel to, or converging distally with, exterior lateral margin.

Mesosoma unusually well developed, considerably wider than head width, and larger in bulk and in length than metasoma. Mesoscutum greatly enlarged, projecting forward in a convexity overhanging pronotum. Scutellum large, convex, nearly as tall as mesoscutum and projecting well above level of propodeum. Propodeum well developed and overhanging petiolar node, posterior propodeal face strongly concave. Forewings short relative to mesosomal length (WI 17–20) and bearing a single submarginal cell. Wing color whitish or yellowish with dark brown veins and stigma. Legs short relative to mesosoma length (FI 37–45).

Petiolar scale taller than node length and bearing a broad crest. Ventral process well developed. Gaster oval in dorsal view, nearly twice as long as broad. Gonostylus produced as a bluntly rounded pilose lobe. Volsella with cuspis present, digitus short and downturned distally.

Dorsal surfaces of body largely devoid of erect setae, occasionally with a few fine, short setae scattered on mesoscutum, scutellum, and posterior abdominal tergites. Venter of gaster with scattered setae. Pubescence dense on body and appendages, becoming sparse only on medial propodeal dorsum.

Color as for worker.

Distribution: Native to the Paraná river drainage of Brazil, Paraguay, Argentina, and Uruguay. Introduced worldwide.

Biology: This important pest species has a literature too extensive to be covered in depth here. An early general review of the biology of this ant is given by Newell and Barber (1913). The spread of Argentine ants around the world is documented by Roura-Pascual et al. (2004), Wild (2004), Giraud et al. (2002), and Suarez et al. (2001). Ecological impacts of Argentine ant invasion have been detailed in numerous studies, including Suarez and Case (2003), Touyama et al. (2003), Christian (2001), and Human and Gordon (1997). Colony structure has also received considerable attention, and relevant papers include Holway and Suarez (2004), Tsutsui and Case (2001), Reuter et al. (2001), and Kreiger and Keller (2000). A series of studies by Cavill and colleagues (Cavill and Houghton 1973, Cavill and Houghton 1974, Cavill et al. 1980) describe some of the glandular and cuticular chemistry of *L. humile*. Chemical and biological control options are reviewed by Harris (2002).

Of the *L. humile* material examined, more than 90% of native range records are within 10 kilometers of a major river in the Paraná drainage. Contrary to some reports (Buczkowski et al. 2004), *L. humile* can reach high densities in urban areas in Argentina and Paraguay (Wild 2004) as well as in less disturbed habitats (Heller 2004). Where nest information was recorded in the native range, 24 nests are from soil, five from under covering objects such as stones or garbage, one from an old termite mound, and one from under bark. This species is polygynous and polydomous, and many nests are recorded as having numerous dealate queens. In contrast to introduced populations, alate queens are not uncommon in nests in Argentina (Wild 2004). One observation in Victoria, Argentina, notes a live lycaenid larva in the brood nest (Wild, pers. obs.).

Similar species: Workers of the sister species *L. oblongum*, from the high Andes of Bolivia and northern Argentina, normally have at least some members of each series with dilute pubescence on gastric tergites 2–4. These ants also have, on average, smaller eyes (OI 28–38, Fig. 86) and longer antennal scapes (SI 120–139, Fig. 85) than *L. humile*. Workers of *L. anathema*, a rarely-collected Brazilian species, have a more produced propodeum (Fig. 33), a narrow head (CI < 86), and usually bear short standing setae on gastric tergites 1–2. Workers of other Humile-group species have shorter antennal scapes and often bear erect setae on the pronotum and basal gastric tergites. Males of related species are much smaller than *L. humile* and lack the greatly swollen mesosoma.

Discussion: The taxonomy and distribution of *L. humile* was recently reviewed in depth by Wild (2004).

Material examined: ARGENTINA. Buenos Aires: Boca, 34°38'S 58°21'W [ALWC]; Buenos Aires, 34°36'S 58°28'W [BMNH, MHNG, NHMB, NHMW, UCDC]; Campana, 34°12'S 58°56'W [ALWC, BMNH, CASC, MZSP]; Reserva Costanera Sur, 34°07'S 58°21'W [AVSC]; Isla Martin Garcia [MACN, NHMB]; La Plata [NHMB]; Lima-Zárete [IFML]; Olivos [MACN]; Reserva Otamendi, 34°14'S 58°54'W [ALWC, AVSC, IFML]; Rosas- F.C.Sud [MACN]; Santa Coloma, 34°26'S 59°02'W [UCDC]. Chubut: Rawson, 43°18'S 65°06'W [PSWC]. Corrientes: Ayo. Cuay Grande, 28°47'S 56°17'W [UCDC]; Corrientes [MACN]; Ita Ibate, 27°25'S 57°10'W [AVSC]; Port Alvear, 29°07'S 56°33'W [AVSC]; Sto. Tomé, 28°33'S 56°03'W [IFML]. Entre Rios: 10 km S Medanos, 33°29'S 58°52'W [ALWC, BMNH]; Colon, 32°15'S 58°07'W [AVSC]; Diamante, 32°01'S 60°39'W [ALWC]; Est. Sosa [MACN, MHNG, NHMB]; Parque Nacional El Palmar, 31°53'S 58°13'W [AVSC]; Parque Nacional Pre Delta, 32°7'S 60°38'W [AVSC]; Port Ibicuy, 33°48'S 59°10' [AVSC]; Victoria, 32°38'S 60°10'W [ALWC]; Villaguay [NHMB]. Formosa: Clorinda [IFML]; Formosa [MACN, NHMB]; Mojon de Fierro [IFML]. La Rioja: Amingá, 28°50'S 66°54'W [ALWC, IFML, UCDC]; Chuquis, 28°54'S 66°58'W [UCDC]. Misiones: Parque Nacional Iguazú, 25°42'S 54°26'W [IFML]; Posadas [MZSP]. Santa Fe: 10 km E Santa Fe, Ruta 168, 31°41'S 60°34'W [ALWC, MCZC, USNM]; Fives Lille [NHMB]; Port Ocampo, 28°30'S 59°16'W [AVSC]; Rosario [MACN]. Tucumán: Tichuco, 26°31'S 65°15'W [UCDC]. AUSTRALIA. New South Wales: Sydney [MZSP]. Victoria (s. loc.) [BMNH]. BELGIUM. Bruxelles Capitale: Brussels [BMNH, NHMB], Brussels Botanical Garden [MHNG]. BERMUDA. Bermuda (s. loc.) [BMNH]. BRAZIL. Amazonas: Manaus [MZSP]. Goiás: Anápolis [MZSP]. Mato Grosso do Sul: Corumbá, Faz. Sta. Blanca [MZSP]; Corumbá, Pto. Esperança [MZSP]; Passo do Lontra, 19°34'S 57°01'W [PSWC, UCDC]; Pto. Murtinho [MZSP]. Rio de Janeiro: Rio de Janeiro [MCSN, MCZC, MHNG]. Rio Grande do Sul: N. Würtemberg [MZSP]; Pelotas [BMNH]. Brazil (s.loc.), Port-of-entry U.S. intercept [ALWC, UCDC, USNM]. CAMEROON. Centre-Sud:

nigellus Emery as infraspecific name in *dispertitus*, based on misidentification of *dispertitus* from Mexico [2 workers, MHNG, examined].

Iridomyrmex melleus Wheeler 1908: 151 (W, Q, M). Lectotype worker, by present designation [MCZC, examined], 55 worker and 1 Queen paralectotypes from Monte Morales, Puerto Rico [MCZC, MHNG, examined]. **syn. nov.**

Iridomyrmex melleus var. *fuscescens* Wheeler 1908: 153 (W). Lectotype worker, by present designation [MCZC, examined], and 5 worker paralectotypes, Mandios, Puerto Rico [MCZC, examined]. **syn. nov.**

Iridomyrmex melleus subsp. *succineus* Forel 1908: 396 (W). Lectotype worker, by present designation [MHNG, examined], and 2 worker paralectotypes, São Paulo, Brazil [MHNG, examined]. **syn. nov.**

Iridomyrmex dispertitus Forel r. *nigellus* Emery. Forel 1912: 47.

Iridomyrmex iniquus r. *succineus* Forel 1908. Forel 1912: 47. First placement of *succineus* as infraspecific name in *iniquum*.

Iridomyrmex iniquus var. *bicolor* Forel 1912: 47 (W). Lectotype worker, by present designation [MHNG, examined], and 6 paralectotypes from Martinique [MHNG, NHMB, examined]. **syn. nov.**

Iridomyrmex melleus var. *dominicensis* Wheeler 1913b: 242 (W). Lectotype worker, by present designation [MCZC, examined], and 10 worker paralectotypes, Roseau, Dominica, F. Lutz [MCZC, examined]. **syn. nov.**

Iridomyrmex melleus st. *succineus* var. *paranaensis* Santschi 1929: 305 (W). Unavailable name. [mat. ref. Rio Negro, Paraná, Brazil, Reichensperger, 23 w, NHMB, examined]

Iridomyrmex iniquus subsp. *nigellus* Emery. Creighton 1950: 341-343. Returned *nigellus* to infraspecific name in *iniquus*.

Iridomyrmex melleus Wheeler. Wheeler, G. C. & J. C. Wheeler 1974: 397. Description of larva.

Linepithema iniquum (Mayr). Shattuck 1992a: 16. First combination in *Linepithema*.

Linepithema iniquum bicolor (Forel). Shattuck 1992a: 16. First combination in *Linepithema*.

Linepithema iniquum nigellum (Emery). Shattuck 1992a: 16. First combination in *Linepithema*.

Linepithema iniquum succineum (Forel). Shattuck 1992a: 16. First combination in *Linepithema*.

Linepithema melleum (Wheeler). Shattuck 1992a: 16. First combination in *Linepithema*.

Linepithema melleum dominicensis (Wheeler). Shattuck 1992a: 16. First combination in *Linepithema*.

Linepithema melleum fuscecens (Wheeler). Shattuck 1992a: 16. First
    combination in *Linepithema.*
Linepithema iniquum (Mayr). Shattuck 1994: 124.
Linepithema iniquum bicolor (Forel). Shattuck 1994: 125.
Linepithema iniquum nigellum (Emery). Shattuck 1994: 125.
Linepithema iniquum succineum (Forel). Shattuck 1994: 125.
Linepithema melleum (Wheeler). Shattuck 1994: 126.
Linepithema melleum dominicensis (Wheeler). Shattuck 1994: 126.
Linepithema melleum fuscecens (Wheeler). Shattuck 1994: 126.
Linepithema iniquum (Mayr). Bolton 1995: 247.
Linepithema iniquum bicolor (Forel). Bolton 1995: 247.
Linepithema iniquum nigellum (Emery). Bolton 1995: 247.
Linepithema iniquum succineum (Forel). Bolton 1995: 247.
Linepithema melleum (Wheeler). Bolton 1995: 247.
Linepithema melleum dominicense (Wheeler). Bolton 1995: 247.
Linepithema melleum fuscecens (Wheeler). Bolton 1995: 247.

Species group: Iniquum

Worker measurements: (n = 100) HL 0.55–0.74, HW 0.46–0.70, MFC 0.13–0.20, SL
0.55–0.78, FL 0.45–0.69, LHT 0.44–0.78, PW 0.29–0.43, ES 1.06–2.03, SI 95–129,
CI 84–97, CDI 24–31, OI 19–29.

Worker diagnosis: Mesonotal dorsum with deep, step-like mesal impression; dorsum
of head posterior of clypeus with > 5 erect setae; mesopleura and metapleural bulla
lacking pubescence and strongly shining.

Worker description: Head in full face view nearly oval in shape, varying from
relatively narrow to relatively broad (CI 84–97), lateral margins convex, posterior
margin convex to straight. Compound eyes of moderate size (OI 19–29), comprised of
50–105 (mean = 67) facets. Antennal scapes long (SI 95–129), slightly shorter to
slightly longer than head length. In full face view, scapes in repose exceeding
posterior margin of head by a length greater than the length of the first funicular
segment. Frontal carinae moderately spaced (CDI 24–31). Maxillary palps long, more
than ½ HL, ultimate segment (segment six) longer than segment 2.

Mesosoma in lateral view relatively slender with dorsal profile comprised of three
distinct convexities: 1) pronotum + anterior mesonotum, 2) posterior mesonotum, 3)
propodeum. Mesonotum with a deep, step-like mesal impression. Mesothorax at
constriction usually narrower than widest diameter of fore coxa. Metanotal groove
deeply impressed. Propodeum raised and varying in profile from evenly rounded to
somewhat flattened dorsally. Metapleural bulla relatively swollen and protruding.

Petiolar scale inclined anteriorly with a relatively blunt node, in lateral view falling short of propodeal spiracle.

Cephalic dorsum (excluding clypeus) bearing 5–13 (mean = 7.7) erect setae, usually arranged in two parallel longitudinal rows extending from the frontal carinae towards the posterior margin. Pronotum bearing 0–7 (mean = 3.4) erect setae, anterior setae usually much longer than posterior setae when posterior setae present. Mesonotum without erect setae. Gastric tergite 1 (= abdominal tergite 3) bearing 3–7 (mean = 4.6) erect setae, tergite 2 with 1–14 (mean = 10.7) erect setae, tergite 3 with 4–11 (mean = 7.0) erect setae. Venter of metasoma with scattered erect setae.

Surface of head and mesosomal dorsum smooth and relatively shining. Pubescence variable, sparse to dense on cephalic dorsum, mesosoma, and gaster, usually more dense on cephalic dorsum than sides of head, and on anterior than posterior gastric tergites. Mesopleura and metapleural bulla always without pubescence, surface glabrous.

Color variable; entire body and appendages pale yellow to brown or black, sometimes bicolored with gaster darker than head, mesosoma, and appendages.

Queen measurements: (n = 6) HL 0.78–0.82, HW 0.73–0.77, SL 0.72–0.76, FL 0.70–0.76, LHT 0.75–0.82, EL 0.23–0.25, MML 1.44–1.63, WL 4.11–4.92, CI 92–97, SI 94–101, OI 30–31, WI 29–30, FI 45–50.

Queen description: Moderately small species (MML < 1.7). Head longer than broad in full face view (CI 92–97), posterior margin straight to slightly convex. Eyes small (OI 30–31). Ocelli small. Antennal scapes moderate to long (SI 94–101), in full face view scapes in repose surpassing posterior margin by a length greater than length of first funicular segment.

Forewings moderately short relative to mesosomal length (WI 29–30). Forewings with Rs+M somewhat longer than M.f2. Legs moderately long relative to mesosomal length (FI 45–50).

Dorsum of mesosoma and metasoma with abundant standing setae. Mesoscutum bearing more than 10 standing setae. Body color variable, from testaceous to piceous.

Male measurements: (n = 5) HL 0.43–0.56, HW 0.41–0.52, SL 0.10–0.15, FL 0.46–0.63, LHT 0.41–0.56, EL 0.19–0.23, MML 0.78–1.01, WL 1.93–2.66, PH 0.17–0.21, CI 91–99, SI 24–26, OI 40–45, WI 23–28, FI 57–62.

Male diagnosis: Forewing with 1 submarginal cell; petiolar node bearing a relatively low, rounded scale; propodeum with posterior face straight to convex; petiolar node concolorous with mesosoma; eye separated from posterolateral margin of clypeus by a distance less than or equal to width of antennal scape.

Male description: Head relatively narrow to about as broad as long in full face view (CI 91–99). Eyes of moderate size (OI 40–45), occupying much of anterolateral surface of head anterior of midpoint and separated from posterolateral clypeal margin by a length less than or equal to width of antennal scape. Ocelli small and in full frontal view emerging only slightly above adjoining posterolateral margins. Antennal scape of moderate length (SI 24–26), about 55–65% length of 3rd antennal segment. Anterior clypeal margin convex medially. Mandibles variable; most commonly mandibles moderately small, masticatory margin slightly longer than inner margin, inner margin and exterior lateral margin parallel to slightly diverging, apical tooth enlarged and followed by a series of 8–12 denticles. More rarely mandibles worker-like, masticatory margin broad and much longer than inner margin, inner margin and exterior lateral margin diverging, apical 2–3 teeth enlarged and followed by a series of 14–18 denticles.

Mesosoma not well developed and subequal in length to metasoma. Mesoscutum not greatly enlarged, not projecting forward over pronotum. Propodeum in lateral view not well developed and not overhanging petiole, posterior face slightly convex and rounding gradually into dorsal face. Forewings of moderate length (WI 23–28) and bearing one submarginal cell. Wings transparent, with yellow, pale brown, or medium brown wing veins and stigma. Legs of moderate length relative to mesosoma (FI 57–62).

Petiolar node bearing a low, rounded scale, node height shorter than node length. Venter of node bearing a convex posteriorly-pointing process. Gaster ovoid in dorsal view, about 2 times as long as broad. Gonostylus produced as triangular pilose lobe. Volsella with cuspis present, digitus short and downturned distally.

Dorsal surfaces of body with erect setae sparse, mesoscutum with 0–10 fine erect to suberect setae. Venter of gaster with scattered setae. Pubescence dense on body and appendages, becoming sparse only on medial propodeal dorsum. Sculpture on head and mesoscutum not well developed, surface shining through pubescence.

Head, mesosoma, and metasoma testaceous to dark brown. Mandibles, antennae, and legs usually lighter than head and mesosoma.

Distribution: Central America, Jamaica, Puerto Rico, the Lesser Antilles, and the northern coast of South America south along the Andes to Paraguay and southeastern Brazil. Incidental worldwide.

Biology: Linepithema iniquum is primarily an arboreal ant. Thirteen nest records are from dead branches or dead wood inside living trees, four are from dead twigs or vines, two records are from the base of a bromeliad, one from a bamboo sheath, two from rotting wood on the ground, and nine nests intercepted in ports-of-entry around the world have been in orchids. Wheeler (1908: 153) reports the ant in Puerto Rico (as *Iridomyrmex melleus*) nesting in hollow twigs, in leaf petioles of plantain, and in "friable carton (earth mixed with vegetable debris) on the under sides of the huge reniform leaves [of *Coccoloba rugosa* (Ortegon)]," and in Dominica "'under and in dead stalks of bananas and plantains'"(Wheeler 1913b: 242). On St. John Island, Pressick and Herbst (1973) report *L. iniquum* nesting in twigs and in logs in moist forest and grassy field habitats. Leuderwaldt (1926) reports two nests in bamboo (as *I. iniquus*), one under bark (as *I. iniquus* var. *succinea*), and one in a dry branch (as *I. iniquus* var. *succinea*) in southern Brazil. *Linepithema iniquum* has been collected rarely in leaf-litter surveys, but these likely reflect incidental ground foragers.

Linepithema iniquum has been collected from sea level to over 2000 meters in elevation. This species is exclusively montane in Central America and northern South America, and is found at varying elevations in the Caribbean, southern Brazil, and Paraguay. 25 museum records are from human disturbed habitats such as 2nd growth forest edges, roadsides, orchards, and pastures. Four are from tropical humid montane forest, seven from primary Atlantic forest in Brazil and Paraguay, three Paraguayan records are Malaise trap samples from low inundated forest, and one Paraguayan record is from an inundated grassland. In Ecuador this species can be locally abundant along roadsides, in pastures, and in *Psidium guayaba* orchards, and in Puerto Rico *L. iniquum* is among the most abundant ants in the mountains (Wheeler 1908, Wild, pers obs).

Linepithema iniquum is probably polydomous, as individual arboreal nests frequently contain no queens (Wild, pers obs). Of five full nest excavations conducted by the author in Ecuador and Puerto Rico that found dealate queens, four colonies had a single dealate queen and one had two dealate queens. This pattern suggests that *L. iniquum* is monogynous to weakly polygynous, although confirmation will require molecular genetic data. Male and female alates have been observed in nests in Brazil from October to April, in Ecuador in August and December, in Costa Rica in November, and in Puerto Rico year round. Both male and female alates have been attracted to lights in November in Puerto Rico.

In Puerto Rico this species has been observed tending *Coccus* and *Saissetia* scale on coffee (Smith 1942: 22), tending pseudococcids on *Cecropia* (Wild, pers. obs), and visiting extra floral nectaries (Wild, pers. obs). There is one observation of aphid-tending in Ecuador (Wild, pers. obs). Smith (1929) and Wheeler (1929) describe the behavior of introduced *L. iniquum* in North American greenhouses in Illinois and Massachusetts, respectively, and observe the ant tending scale, visiting flowers, and nesting in and under soil pots.

Like its better-known congener *L. humile*, *L. iniquum* is carried around the world with human commerce (Fig. 108). More than a dozen museum records of this species are intercepts at various ports-of-entry and quarantine in the United States and Europe, usually carried with epiphytes. Unlike *L. humile*, this species apparently has not been successful in establishing outdoor populations in spite of the opportunity to do so, although there are a few records of this ant persisting in greenhouses in the temperate zone (Wheeler 1929, Creighton 1950).

Similar species: Workers of the closely-related *L. dispertitum* are structurally similar but the mesonotal impression is normally less well developed (Fig. 21), the scapes often somewhat shorter (Fig. 82), and in most populations have fewer than 4 standing setae on the dorsum of the head posterior to the clypeus. In the region where the two species co-occur, from Guatemala to Panamá, *L. iniquum* often shows a reduction in the development of the mesonotal impression such that it occasionally approaches the condition of *L. dispertitum*. However, the two species in this region can usually be separated by differences in pilosity and pubescence. *Linepithema dispertitum*, while pilose in parts of northern Mexico, lacks the extensive pilosity of *L. iniquum* in southern Central America, having four or fewer erect setae on the head, and always bearing dense pubescence on gastric tergites 1 and 2. *Linepithema dispertitum* also shows dense gastric pubescence on gastric tergites 1–2 (sometimes dilute on 3–4), while *L. iniquum* collections from this region nearly always have sparse gastric pubescence on tergites 2–4. Additionally, *L. dispertitum* frequently nests in soil or rotting wood, while *L. iniquum* is normally arboreal. Males of *L. leucomelas* are nearly indistinguishable from *L. iniquum* males, but the petiolar node of *L. leucomelas* males is much lighter in color than the mesosomal dorsum.

Discussion: *L. iniquum* exhibits considerable variation among the worker caste in size, color, propodeal form, pilosity, and pubescence, which has lead to a proliferation of names. I find little justification in retaining the multiplicity of names associated with *L. iniquum* and *L. melleum*. Most of these names describe local color forms, and these forms inevitably lose their distinctness with the examination of sufficient quantity of material. Much of the variation is allopatric. The few instances of sympatric variation are normally with respect to only one character, and usually the character is present in an intermediate state either in the same location or within specimens collected within a few hundred kilometers. Males are remarkably similar in form and in genital structure across the range, differing primarily in color and more rarely in mesonotal pilosity. The arboreal nesting habit is also apparently uniform across the range.

Specimens from the eastern Andean slopes in Ecuador, Peru, and parts of Colombia, as well as from Jamaica and some high elevations in Central America, are nearly always medium brown to black in color and have sparse pubescence on gastric tergites 2–4. Mayr's *Hypoclinea iniqua* holotype from Bogota is damaged, with only a gaster and a few tarsi remaining, but it unambiguously belongs to this form as the gaster has the distinctive sparse pubescence and dark coloration. Costa Rican specimens, including Emery's types of *I. iniquus nigellus*, are often more robust than most other *L. iniquum*, with a less pronounced mesonotal impression, but these are otherwise similar to other dark sparsely pubescent forms. Specimens from the western Andean slopes from Ecuador north through Calí, Colombia, as well as from Guyana north through the Lesser Antilles to Puerto Rico, have moderate to dense pubescence on all gastric tergites. These pubescent forms are variable in color, from yellow to dark brown, without much geographic structure to the color variation. Puerto Rican specimens are most commonly pale yellow, as in Wheeler's types of *I. melleus*, but there is continuous variation to bicolored or to a medium brown, as in Wheeler's *I. melleus fuscecens*. Wheeler's *I. melleus dominicensis* types from Dominica appear indistinguishable from paler specimens found in Puerto Rico. Venezuelan and northern Colombian material may be of either pubescence form or intermediate, with both forms recorded in sympatry in Valle de Cauca, Colombia. Forel's *I. iniquus bicolor* types from Martinique are sparsely pubescent and sharply bicolored with a yellow head and mesosoma and a dark gaster.

Atlantic forest material from Brazil presents a more confusing array of variation, as specimens can be small to large, with scapes of varying length, pale yellow to black or bicolored, and sparsely to densely pubescent, with little apparent geographic pattern to the variation and occasional local sympatry of more than one form. A dark densely pubescent form, a yellow sparsely pubescent form, and intermediates have been recorded in sympatry at Nova Teutônia, Santa Catarina. A dark sparsely pubescent form, a yellow densely pubescent form, and a bicolored form are all recorded from Rio de Janeiro and São Paulo states, sometimes in local sympatry. Many specimens from Paraná state are bicolored. Paraguayan material is dark brown, sparsely pubescent, with relatively small eyes (50–60 ommatidia).

Forel's types of *I. iniquus succineus* from São Paulo, as well as several other collections from São Paulo and Rio de Janeiro states, are distinct in the combination of large size, dense gastric pubescence, and yellow-orange coloration. Further study into the ecology and genetics of this form may reveal that *L. succineum* merits elevation to specific status (conceivably rendering the remainder of *L. iniquum* paraphyletic). However, given that each of the character states that in combination define this form can be found individually or in other combinations among other southern Brazilian collections of *L. iniquum*, I find it preferable to retain *L. succineum* as a synonym of *L. iniquum*.

Material examined: AUSTRIA. Steiermark: Graz [MHNG]. BELGIUM. Bruxelles, Jardin Botanico [NHMB]; Bruxelles [BMNH]. BRAZIL. Bahia: Maracás, Faz. Maria Inácia [MZSP]. Espírito Santo: 4 km N Santa Teresa, Reserva Nova Lombardia, 900m [MCZC]. Minas Gerais: Serra Caraça, 1380m [MZSP]. Paraná: Pq. Marumby km 34, Estr. Graciosa [MZSP]; Pto. Vitória [MZSP]; Rio Negro [MZSP, NHMB]; Taquara [MZSP]. Rio de Janeiro: Ilha da Marambaia [MZSP, UCDC]; Itatiaia [MZSP]; Petrópolis [MCZC, MZSP, NHMB]; Petrópolis, Tr. de Fatima [MZSP]. Rio Grande do Sul: Cotegipe [MZSP]; Rio Grande do Sul (s.loc.) [MCSN, MHNW]. Santa Catarina: Bal. Camboriu [MZSP]; Blumenau [MHNG, MZSP]; Gaspar [MZSP]; N. Teutônia, 27°11'S 052°23'W [MZSP]; Rodeio [MZSP]; Santa Catarina (s. loc.) [BMNH, MHNW]. São Paulo: Agudos [MZSP]; Alto da Serra [MZSP]; Botucatu [MHNG]; Cachoeira das Emas, Mun. Piraçununga [MCZC]; Embu [MZSP]; Guarantingueta [MZSP]; Ilha dos Buzios [MZSP]; Ipirangá [MZSP]; Salesópolis, Est. Biol. Boraceia [MZSP]; Sao Paulo [MCSN, MHNG]. Brazil (s.loc.) [MCSN, MCZC, MHNG]. COLOMBIA. Antioquia: Medellín,1800m [MCZC]. Cundinamarca: Cáqueza [WPMC]. La Guajira: San Antonio [MHNG]. Quindío: Armenia [WPMC]. Risaralda: Apia, La Estrella, 1470m [ALWC]. Valle de Cauca: Alcala [ALWC]; Calí [MZSP]; Colima Site [WPMC]. Colombia (s. loc.), Port-of-entry U.S. intercept [MCZC, MZSP]. COSTA RICA. Cartago: Cartago [MCZC]; 2 km N Cervantes, 1600m [MCZC]; Guatuso, 450m [MHNG]; Tres Ríos, San José [MHNW]. Heredia: 6 km ENE Vara Blanca, 2000m [JTLC]. San José: San José [MCSN, MCZC]; Santa Maria, 1600m [MHNG]. Costa Rica (s.loc.) [MCSN, MHNG, NHMB]. DOMINICA. St. George: Roseau [MCZC]. ECUADOR. Pichincha: ENDESA forest concession [MCZC]; Las Pampas-Alluriquin Road, 00°21'S 79°00'W, 970m [ALWC, LACM]; Maquipucuna, 5 km ESE Nanegal, 00°07'N 078°38'W, 1250m [MCZC, PSWC, UCDC]; Road to Mindo, 00°03'S 078°47'W, 1325m [ALWC, LACM, MZSP, QCAZ]; Nanegalito, 00°04'N 78°40'W, 1450m [ALWC, BMNH, CASC, IFML, JTLC, LACM, MCZC, MHNG, MZSP, NHMB, NHMW, PSWS, QCAZ, USNM, WPMC]; Unión Toachi, 00°19'S 78° 57'W, 1000m [ALWC, BMNH, MCZC, MZSP]. Tungurahua: Baños, 01°24'S 78°25'W, 1900m [ALWC, BMNH, LACM, MCZC, MZSP, QCAZ]. Ecuador (s. loc.), Port-of-entry U.S. intercept [USNM]. EL SALVADOR. Boqueron to Quetzaltepeque, 1600m [MCZC]. GERMANY. Brandenburg: Potsdam [MCZC]. GRENADA. Saint Andrew: Grand Etang [LACM, MCZC]. Saint Patrick: Sauteurs [MCZC]. GUATEMALA. Guatemala (s.loc.), Port-of-entry U.S. intercept [LACM, USNM]. IRELAND. Dublin: Dublin [BMNH]. JAMAICA. Kingston-St. John: Kingston [MCZC]; Kingston Botanical Gardens [MHNG]. Manchester: Mandeville [MCZC]. Portland: 1 km W Section, 18°05'N

76°42'W, 990m [ALWC, CASC, MZSP]. Saint Andrew: Cinchona, 18°04'N 76°39'W 1450m [ALWC, BMNH, IFML, JTLC, LACM, MCZC, UCDC]. Saint Ann: Cedar V., 760m [LACM]; Welcome: 18°17'N 77°20'W, 690m [JTLC, LACM]. Saint James: Great R., 18°26'N 77°59'W 0-80m [JTLC]. Trelawny: Quick Step, 18°18'N 77°45'W, 380m [LACM]. Jamaica (s.loc.) [MCZC]. MARTINIQUE. Martinique (s.loc.) [MHNG, NHMB]. MEXICO. Mexico-Guatemala (s.loc.), Port-of-entry U.S. intercept [MCZC, USNM]. PANAMA. Panamá: Culebra, c.z. [MCZC]. PARAGUAY. Canindeyú: Reserva Mbaracayú, Karapá, 24°00'S 55°19'W, 500m [ALWC, LACM]; Reserva Mbaracayú, Jejuimí, 24°08'S 55°32'W, 170m [ALWC, BMNH, CASC, INBP, LACM, MCZC, MZSP, UCDC]. Cordillera: San Bernadino [MHNG, NHMB]. Itapua: Cap. Mesa [MZSP]. PERU. Cusco: 80-100 km NW Cusco, 910m [USNM]; Machu Picchu, 2000-2200m [MCZC]. Pasco:3 km N Oxapampa, 2000m [LACM]. SAINT VINCENT AND THE GRENADINES. St. Vincent: Soufriere Vol., 910m [LACM]. St. Vincent (s,loc.) [MHNG]. USA. Massachusetts: Forest Hills, Bussey greenhouse [LACM, MCZC]. Puerto Rico: Doña Juana Recreational Area [WPMC]; Humacao, El Yunque [ALWC, LACM, MCZC, MZSP, WPMC]; Carribean National Forest (El Yunque), Las Cabezas [ALWC, LACM]; Humacao, El Verde, 130m [LACM]; Humacao, carr.184 Res. For. El Carite [LACM]; Humacao, carr. 7740, ca 10 km SSW San Lorenzo, 550m [LACM]; Humacao, Mt. Britton, 940m [LACM]; Humacao, Valle de Icacos, 700m [LACM]; Indeira [MZSP]; Rio Grande, El Verde Field Station [WPMC]; Bo. Sonadora, Guaynabo [LACM]; Between Salinas & Cayay [LACM]; Carite Forest Reserve, 500m [ALWC, CASC, LACM]; Maricao Forest Reserve, 600–750m [ALWC, BMNH, LACM, MCZC, MZSP, USNM]; Isabela [LACM]; Mandios [MCZC]; Mayaguez [BMNH, LACM]; Monte Morales [MCZC, MHNG]; Mona Is. [MCZC]; Naguabo [MCZC]; Rio Abajo Forest Reserve, 330m [ALWC]; Rio Grande [LACM]; Toro Negro Forest, nr. Manatí [LACM]; Utuado [LACM, MCZC]. US Virgin Islands: Bordeaux Mt., St. John [MCZC]; Mt. Eagle [NHMB]; St. John [LACM]. VENEZUELA. Aragua: Rancho Grande [MZSP]. Venezuela (s.loc.), Port-of-entry U.S. intercept [LACM, MZSP, USNM].

Linepithema keiteli (Forel)

(worker mesosoma Fig. 15; worker head Fig. 16; male head Fig. 59; male volsella Fig. 60; distribution Fig. 107)

Iridomyrmex keiteli Forel 1907: 8 (W, Q). Lectotype worker, by present designation [MHNG, examined], and 3 worker paralectotypes, Port-au-Prince, Haiti, G. Keitel, 6.xii.1901, [MCZC, MHNG, examined].
Iridomyrmex keiteli var. *subfasciatus* Wheeler and Mann 1914: 43 (W).

Lectotype worker, by present designation [MCZC, examined], and 2 worker paralectotypes, Petionville, Haiti, W. M. Mann [MCZC, examined]. **syn. nov.**

Linepithema keiteli (Forel). Shattuck 1992a: 16. First combination in *Linepithema*.

Linepithema keiteli subfasciatum (Wheeler and Mann). Shattuck 1992a: 16. First combination in *Linepithema*.

Linepithema Keiteli (Forel). Shattuck 1994: 125. [incorrect subsequent spelling.]

Linepithema keiteli subfasciatum (Wheeler and Mann). Shattuck 1994: 125.

Linepithema keiteli (Forel). Bolton 1995: 247.

Linepithema keiteli subfasciatum (Wheeler and Mann). Bolton 1995: 247.

Species group: Fuscum

Worker measurements: (n = 24) HL 0.59–0.77, HW 0.55–0.75, MFC 0.14–0.20, SL 0.54–0.69, FL 0.45–0.63, LHT 0.47–0.67, PW 0.37–0.48, ES 1.08–1.99, SI 91–103, CI 92–101, CDI 26–30, OI 17–28.

Worker diagnosis: A robust species with a relatively broad head (CI 92–101); antennal scapes in repose surpass posterior margin of head by a distance about equal to length of first funicular segment; mesonotum with sparse medial pubescence, surface smooth and strongly shining; gastric tergites 1–2 each with > 10 erect or suberect setae.

Worker description: Head in full face view slightly longer than broad to about as long as broad (CI 92–101), lateral margins convex, posterior margin slightly to strongly concave. Compound eyes small to moderate size (OI 17–28), comprised of 40–75 ommatidia. Antennal scapes shorter than head length to approximately as long as head length (SI 91–103). In full face view, scapes in repose surpass posterior margin of head by a distance about equal to length of first funicular segment. Frontal carinae moderately spaced (CDI 26–30). Maxillary palps of moderate length, slightly longer than ½ HL, ultimate segment (segment six) subequal in length to segment 2.

Mesosoma in lateral view with pronotum and mesonotum forming a single continuous convexity, mesonotum strongly convex, without central saddle or indentation, in larger workers meeting dorsal propodeal face at an angle of 90°–110°. Metanotal groove slightly impressed, propodeum depressed well below level of mesonotum. Propodeum in lateral view rounded, not sloping forward, posterior margin convex dorsad of the metapleural bulla. Metapleural bulla protruding.

Petiolar scale inclined anteriorly, in lateral view falling short of propodeal spiracle.

Cephalic dorsum (excluding clypeus) bearing 4–9 erect setae. Pronotum bearing 2–10 (mean = 5.7) erect to suberect setae of varying length. Mesonotum without erect setae, or rarely with a single short suberect seta. Gaster pilose, setae somewhat irregular in length and inclination, shorter setae often difficult to distinguish from pubescence, gastric tergite 1 bearing 10–21 erect setae (mean = 14.8), tergite 2 with 11–26 erect setae (mean = 16.2). Venter of metasoma with scattered erect setae.

Sculpture on head and pronotal dorsum lightly reticulate-punctate and strongly shining. Pubescence dense on dorsum of head, anterior petiolar scale, and gastric tergites 1–4. Pubescence relatively long and subdecumbent. Pubescence on head often fading to sparse laterally. Mesonotal dorsum with moderate to sparse pubescence fading to sparse medially, surface smooth and strongly shining. Lateral face of pronotum, mesopleura and metapleural bulla without pubescence and strongly shining.

Color variable, from head and mesosoma piceous with gaster and appendages medium brown to head and mesosoma light reddish brown with gaster and appendages pale yellow. Lateral clypeal extensions and genae surrounding mandibular insertions lighter in color than remainder of head. Gaster usually lighter than head, tarsi and trochanters lighter than femora and tibiae.

Queen measurements: (n = 3) HL 0.93–1.02, HW 0.93–1.02, SL 0.83–0.91, FL 0.89–0.97, LHT 1.01–1.16, EL 0.29–0.32, MML 1.88–2.15, WL 5.98–7.07, CI 100–103, SI 88–90, OI 31–32, WI 32–33, FI 43–47.

Queen description: Relatively large species (MML 1.88–2.15). Head as broad or broader than long in full face view (CI 100–103), posterior margin slightly convex. Eyes moderately small (OI 31–32). Ocelli of moderate size. Antennal scapes of moderate length (SI 88–90), in full face view scapes in repose surpassing posterior margin by a length about that of first funicular segment.

Forewings moderately long relative to mesosomal length (WI 32–33). Forewings with Rs+M somewhat shorter to somewhat longer than M.f2. Legs moderate relative to mesosomal length (FI 43–47).

Dorsum of mesosoma and metasoma with abundant fine suberect to subdecumbent setae, mesoscutum with more than 20 subdecumbent setae. Body, femora, and often antennal scapes dark brown. Tibiae, tarsi, and trochanters lighter in color.

Male measurements: (n = 4) HL 0.61–0.70, HW 0.56–0.65, SL 0.21–0.22, FL 0. 88–1.06, LHT 0.85–1.10, EL 0.24–0.28, MML 1.31–1.61, WL 3.66–4.68, PH 0.32–0.33, CI 85–98, SI 31–34, OI 37–42, WI 28–29, FI 66–68.

Male diagnosis: Forewing with 2 submarginal cells; eyes relatively small (OI < 43); volsella with distal arm greater than 2/3 length of proximal arm and much longer than height of volsella in lateral view.

Male description: Head slightly longer than broad in full face view (CI 94–98). Eyes relatively small (OI 37–42), occupying about 1/3 lateral surface of head at midlength and separated from posterolateral clypeal margin by a length less than or equal to width of antennal scape. Ocelli of moderate size and in full frontal view set above adjoining posterolateral margins. Antennal scape moderately long (SI 31–34), 85–100% length of 3rd antennal segment. Anterior clypeal margin broadly convex medially. Mandibles large and nearly worker-like, masticatory margin broad, much longer than inner margin, bearing 1–2 apical teeth and 10–14 denticles. Inner margin and exterior lateral margin diverging.

Mesosoma moderately developed, shorter in length than metasoma. Mesoscutum slightly enlarged, not projecting strongly forward or overhanging pronotum. Scutellum large, convex, nearly as tall as mesoscutum and projecting well above level of propodeum. Propodeum in lateral view not overhanging petiole, dorsal face rounding evenly into posterior face, posterior face straight to convex. Forewings long relative to mesosomal length (WI 28–29) and bearing two submarginal cells. Wing color clear to slightly smoky, with darker brown veins and stigma. Legs long relative to mesosoma length (FI 66–68).

Petiolar node bearing a blunt, broadly-rounded scale, node height less than node length. Ventral profile of node convex. Gaster elongate in dorsal view, 2.5–3 times as long as broad. Gonostylus produced as a slender filament. Volsella with ventrodistal process present as a sharp tooth. Cuspis absent. Digitus elongate, distal arm long, at least 2/3 length of proximal arm, and much longer than height of volsella in lateral view. Proximal arm gradually narrowed from ½ height of volsella at base to filamentous at juncture with distal arm.

Dorsal surfaces of body with scattered erect setae, more than 10 erect setae on mesoscutum. Venter of gaster with scattered setae. Pubescence dense on body and appendages, becoming sparse only on medial propodeal dorsum. Head somewhat shining through pubescence.

Head usually dark brown, darker than body. Body and appendages medium brown. Mandibles, trochanters, tarsi, and terminal antennal segments lighter.

Distribution: Mountains of Hispaniola.

Biology: This species is among the most abundant ants in the mountains of Hispaniola. Collections run from 770 to 1700 meters in elevation. Most records are from montane pine forest down through mesic tropical moist forests, and several collections were made on deforested slopes, forest edges, and roadsides. 21 nesting records are from under stones and four from open soil. Two records are from sifted litter. Colonies are populous and form extensive foraging trails during both day and night. Nests often have several entrances and may extend under series of stones. Of seven nest excavations I conducted in the Sierra Baoruco and the Cordillera Central in November 2002, single dealate queens were found in two of the nests, the rest contained only alate queens or none. This observation suggests that *L. keiteli* is monogynous and polydomous, although confirmation will require molecular genetic data. One colony was observed tended aphids in a *Psidium guayaba* tree. *L. keiteli* workers recruit to dead insects and were observed in the Sierra Baoruco attacking a live dipteran larva.

Alate males and queens have been taken from nests in November, and on one occasion in early November 2002 I observed a partially aborted mating flight at dusk. At 18:00hr, males and workers were observed clustering at several nest entrances of a single colony. By 18:30, males began running about excitedly, climbing grass blades. Workers mostly restrained the males by grabbing the males' wings, legs, and genitalia with their mandibles, but at least five males escaped and flew straight upward. By 19:30, the activity had subsided and the remaining males were no longer agitated. No queens were observed.

Similar species: Workers of *L. flavescens*, another Hispaniolan endemic, are closely similar in structure but have smaller eyes (< 35 ommatidia), shorter scapes (SI < 85, Fig. 79), and are pale yellow in color. Workers of Fuscum-group species on the mainland, *L. piliferum*, *L. tsachila*, *L. fuscum* and *L. angulatum*, have a punctuate-reticulate sculpture on the head and mesosoma and are less shining. Males of these species have relatively larger eyes (OI > 43, Fig. 89) than *L. keiteli* males.

Discussion: This species exhibits considerable allopatric variation. Worker specimens from the Cordillera Central tend to be darker with larger eyes, slightly longer scapes, and slightly shorter pubescence than Forel's Haitian types and specimens from Haiti and Sierra Baoruco. Wheeler and Mann's *L. keiteli subfasciatum* from Haiti, while somewhat darker than Forel's *keiteli* types, falls well within the variation of this species and is synonymized here.

Given the variability of *L. keiteli*, the recognition of Wheeler and Mann's *L. flavescens* as a separate species may render *L. keiteli* paraphyletic. However, the morphometric variation among populations of *L. keiteli* is largely continuous, while the variation between *L. keiteli* and *L. flavescens* is marked by discrete differences in antennal scape length, eye size, and coloration.

Material examined: DOMINICAN REPUBLIC. La Vega: 1 km E Reserva Cientifica Ebano Verde, 19°02'N 70° 29'W, 940m [ALWC, BMNH, MCZC]; 12 km NW Bonao, 19°02'N 70°29'W, 890m [ALWC, JTLC]; 14 km NW Bonao, 19°02'N 70°30'W, 1100m [PSWC]; 3 km E El Rio, 19°00'N 70°35'W, 1000m [ALWC, BMNH, LACM, NHMW, QCAZ, USNM]; 3 km N El Rio, 19°00'N 70°38'W, 1270m [ALWC]; 5 km SE Constanza, 18°52'N 70°41'W, 1520m [ALWC, LACM, UCDC]; Casabito Forest, between El Rio and Bonao km 17, 1525m [MCZC]; Casabito Forest, between El Rio and Bonao km 18, 1400m [MCZC]; Cervantia 18°51'N 70°42'W, 1730m [ALWC, IFML]; 5.5 km S Constanza, Road S Balneario, R. Grande, 1500m [MCZC]. Pedernales: 16 km ENE Pedernales, 18°07'N 71°37'W, 800m [PSWC]; 22 km E Pedernales, 18°09'N 71°35'W, 1600m [PSWC]; P. N. Sierra Baoruco, 18°09'N 71°35'W, 1580m [ALWC, LACM, MCZC, MZSP]; P. N. Sierra Baoruco, 18°07'N 71°37'W, 770m [ALWC, CASC, NHMB]; P. N. Sierra Baoruco, Charca, 18°07'N 71°36'W, 1130m [ALWC, BMNH, MZSP, USNM]; P. N. Sierra Baoruco, Las Abejas, 18°09'N 71°37'W, 1320m [ALWC, MHNG]; Sierra de Baoruco, N Cabo Rojo, km 28, 1200m [MCZC]. Santiago: P.N. Armando Bermudez, Los Tablones, 19°03'N 70°53'W, 1300m [ALWC, CASC, MCZC, PSWC, UCDC]. HAITI: Ouest: Diquini [MCZC]; Petionville [MCZC]; Port-au-Prince [MCZC, MHNG]. Sud-Est: Mountains N of Jacmel [MCZC].

Linepithema leucomelas (Emery)

(worker mesosoma Fig. 25; worker head Fig. 26; male head Fig. 67; distribution Fig. 100)

Iridomyrmex leucomelas Emery 1894b: 378 (W). Lectotype worker, by
      present designation [USNM, examined], and 1 worker paralectotype,
      Rio Grande do Sul, Brazil [USNM, examined].
Iridomyrmex leucomelas Emery. Kempf 1969: 294 (W). Rediagnosis of
      worker. *Iridomyrmex aspidocoptus* Kempf 1969: 295-296 (W).
      Holotype worker, Moji das Cruzes, São Paulo, Brazil, 1926, Eschrich
      [MZSP, examined]; 3 paratype workers, Iporanga, Brazil, 1.xi.1961,
      Lenko & Reichardt [MZSP, examined]. **syn. nov.**
Linepithema leucomelas (Emery). Shattuck 1992a: 16. First combination in
      Linepithema.
Linepithema aspidocoptum (Kempf). Shattuck 1992a: 16. First combination in
      Linepithema.
Linepithema aspidocoptum (Kempf). Shattuck 1994: 122.

Linepithema leucomelas (Emery). Shattuck 1994: 126.
Linepithema aspidocoptum (Kempf). Bolton 1995: 247.
Linepithema leucomelas (Emery). Bolton 1995: 247.

Species group: Iniquum

Lectotype worker measurements: HL 0.69, HW 0.60, MFC 0.16, SL 0.70, FL 0.60, LHT 0.61, PW 0.37, ES 1.79, SI 116, CI 88, CDI 27, OI 26.

Worker measurements: (n = 22) HL 0.60–0.78, HW 0.52–0.74, MFC 0.14–0.18, SL 0.64–0.79, FL 0.51–0.68, LHT 0.58–0.71, PW 0.33–0.58, ES 1.22–2.00, SI 106–125, CI 85–96, CDI 24–28, OI 20–26.

Worker diagnosis: Strikingly bicolored, dark head contrasting with pale whitish yellow on all or part of the mesosoma and appendages; antennal scapes long (SI 106–125).

Worker description: Head in full face view varying from relatively narrow to relatively broad (CI 85–96), lateral margins convex, posterior margin slightly convex to concave. Compound eyes of moderate size (OI 20–26), comprised of 55–80 facets. Antennal scapes long (SI 106–125), as long or longer than head length. In full face view, scapes in repose exceeding posterior margin of head by a length greater than the length of the first funicular segment. Frontal carinae narrowly to moderately spaced (CDI 24–28). Maxillary palps very long, well more than ½ HL, ultimate segment (segment six) longer than segment 2.

Mesosoma in lateral view with pronotum and mesonotum forming a single continuous convexity, mesonotal profile straight to very slightly impressed mesally. Metanotal groove slightly impressed to not at all impressed, propodeum low and quadrate-forward.

Petiolar scale sharp and inclined anteriorly, in lateral view falling short of the propodeal spiracle.

Cephalic dorsum (excluding clypeus) bearing 4–6 erect setae. Pronotum bearing 1–4 erect setae. Mesonotum without erect setae. Gastric tergite 1 bearing 1–3 erect setae (mean = 2.0), tergite 2 with 2–4 erect setae, tergite 3 with 4–6 erect setae. Venter of metasoma with scattered erect setae.

Surface of head and mesosomal dorsum finely reticulate-punctate and relatively dull. Pubescence dense on head and mesosomal dorsum. Mesopleura with pubescence fading to absent anteroventrally, surface smooth and shining. Metapleural bulla with moderate to sparse pubescence thinner than on propodeal dorsum but never completely absent. Gastric tergites 1–4 with dense pubescence, surface only slightly shining.

Body color variable, but always bicolored with head and part of gaster dark brown and with at least pronotum, petiole, and some appendages a pale whitish-yellow. Some specimens pale whitish yellow except for dark head and gastric infuscation, others dark except for light pronotum, petiole, funiculus, tibiae and tarsi. Most specimens show mesopleural infuscation.

Queen measurements: (n = 2) HL 0.85, HW 0.78–0.79, SL 0.79–0.83, FL 0.78–0.80, LHT 0.88, EL 0.26–0.27, MML 1.65–1.66, WL 5.01, CI 92–94, SI 100–106, OI 31–32, WI 30, FI 47–49.

Queen description: Moderately small species (MML < 1.7). Head longer than broad in full face view (CI 92–94), posterior margin slightly concave to slightly convex. Eyes small (OI 31–32). Ocelli small. Antennal scapes long (SI 100–106), in full face view scapes in repose surpassing posterior margin by a length greater than length of first funicular segment.

Forewings of moderate length relative to mesosomal length (WI 30). Forewings with Rs+M somewhat longer than M.f2. Legs moderately long relative to mesosomal length (FI 47–49).

Dorsum of mesosoma and metasoma with abundant standing setae. Mesoscutum bearing more than 10 standing setae. Head, mesosoma and metasoma medium to dark brown, trochanters, tibiae, and tarsi testaceous.

Male measurements: (n = 2) HL 0.48–0.52, HW 0.44–0.47, SL 0.11–0.12, FL 0.52–0.55, LHT 0.49–0.54, EL 0.19–0.20, MML 0.87–0.94, WL 2.23, PH 0.19–0.21, CI 91–92, SI 23, OI 38–41, WI 26, FI 58–59.

Male diagnosis: Forewing with 1 submarginal cell; petiolar node bearing a relatively low, rounded scale; propodeum with posterior face convex; antennal segment 3 more than twice as long as antennal scape; dorsum of petiole much lighter in color than mesosomal dorsum.

Male description: Head relatively narrow in full face view (CI 91–92), lateral margins continuing to expand posterior of compound eyes. Eyes moderately small (OI 38–41), occupying much of anterolateral surface of head anterior of midpoint and separated from posterolateral clypeal margin by a length less than width of antennal scape. Ocelli small and in full frontal view emerging only slightly above adjoining posterolateral margins. Antennal scape short (SI 23), only 40–50% length of 3rd antennal segment. Anterior clypeal margin convex medially. Mandibles small, masticatory margin slightly longer than inner margin, apical tooth enlarged and followed by a series of 8–12 denticles. Inner margin and exterior lateral margin parallel to slightly diverging.

Mesosoma not well developed and subequal in length to metasoma. Mesoscutum not greatly enlarged, not projecting forward over pronotum. Propodeum in lateral view not well developed and not overhanging petiole, posterior face slightly convex and rounding gradually into dorsal face. Forewings of moderate length (WI 26) and bearing one submarginal cell. Wings transparent, with pale brown wing veins and stigma. Legs of moderate length relative to mesosoma (FI 58–59).

Petiolar node bearing a low, rounded scale, node height shorter than node length. Venter of node bearing a convex posteriorly-pointing process. Gaster ovoid in dorsal view, about 2 times as long as broad. Gonostylus produced as triangular pilose lobe. Volsella with cuspis present, digitus short and downturned distally.

Dorsal surfaces of body with erect setae sparse, mesoscutum with 0–4 fine erect to suberect setae. Venter of gaster with scattered setae. Pubescence dense on body and appendages, becoming sparse only on medial propodeal dorsum. Sculpture on head and mesoscutum not well developed, surface shining through pubescence.

Head, mesosoma, and metasoma medium brown. Mandibles, petiole, antennae, and legs pale whitish-yellow to light brown, much lighter than head and mesosoma.

Distribution: Southeastern Brazil.

Biology: This species is an Atlantic forest ant. Most records are from rainforest or wet forest. *Linepithema leucomelas* is one of only two primarily arboreal species in the genus, the other being *L. iniquum*. Kempf (1969: 294) observes that *L. leucomelas* nests "under bark, also in the cavities of bamboo, gourd trees, and arboreal ferns." Leuderwaldt (1926) reports several nests from bamboo and one from a *Hemitelia* fern. Ketterl et al. (2003) note *L. leucomelas* nesting in bromeliads in *Araucaria* trees in Rio Grande do Sul. Among museum records, one nest is from under *Jacaranda* bark and two separate Port-of-Entry intercepts into the United States report the ant from orchid plants. Alate queens have been collected from nests in May and October, and alate males have been collected in October.

As noted in Emery's original description and again by Kempf (1969), this ant shows similar coloration to an unrelated arboreal Atlantic forest species, *Tapinoma atriceps* Emery. The significance of this convergence is unknown.

Similar species: *Linepithema leucomelas* is unlikely to be confused with any other *Linepithema* species owing to its striking coloration. Workers of the closely-related *L. iniquum* have a strong constriction of the mesonotum (Fig. 23), while males of *L. iniquum* are nearly identical but the petiolar node is concolorous with the mesosomal dorsum.

Discussion: Some worker specimens from Santa Catarina are darker in color than specimens from elsewhere, with a dark propodeum and more extensive infuscation. A few specimens from São Paulo state are larger and bear more than one pair of pronotal setae. Borgmeier's *aspidocoptum* types from São Paulo are unusual in their broad heads, large size, more extensive pilosity, longer pubescence, and lack of mesopleural infuscation. However, all of these character states except the pubescence are present in various combinations in other specimens of *L. leucomelas*. Once the full variation encompassed by *L. leucomelas* is considered, there is little reason for retaining *L. aspidocoptum* as a separate species.

Material examined: BRAZIL. Minas Gerais: Pq. Est. de Ibitipoca [MZSP]; Serra Caraça (Engenho) [MZSP]. Paraná: Capão Imbuia, Curitíba [MCZC]; Rio Azul [MZSP]. Rio de Janeiro: Col. Alpina [MHNG]; Petrópolis [MZSP]. Rio Grande do Sul: Erechim, Campinas [MZSP]; Rio Grande do Sul (s.loc.) [USNM]. Santa Catarina: Concórdia Estr. Uruguai [MZSP]; Ibirama [MZSP]; Ituporanga [MZSP]; Lanca [MZSP]; N. Teutônia [MZSP]; Queçaba [MZSP]; Seara, 24°07'S 52°18'W [MZSP]; Santa Catarina (s. loc.) [MHNG, MCSN, NHMW]. São Paulo: Agudos [MZSP]; Alto da Serra [NHMB]; Cantareira, SP Capitol [MZSP, NHMW]; Caraguatatuba, Res. Flor. [MZSP, MCZC]; Guarantingueta [MZSP]; Ilha dos Buzios [MZSP]; Iporanga [MZSP]; Jardim Botanico (Agua Funda) [MCZC]; Moji das Cruzes [MZSP]; Salesópolis, Est. Biol. Boraceia [MZSP]; Sao Paulo [MHNG]; S. Sebastião, Barra de Una [MZSP]. Brazil (s. loc.), Port-of-entry U.S. intercept [LACM, NHMB, USNM].

Linepithema micans (Forel), **stat. nov.**

(worker mesosoma Figs. 5, 37, 39; worker head Figs. 38, 40; male head Fig. 76; male lateral Fig. 52; distribution Fig. 103)

Iridomyrmex dispertitus subsp. *micans* Forel 1908: 394-395 (W, M).
Lectotype worker, by present designation [MHNG, examined], 2 worker and 1 male paralectotypes, São Paulo, Brazil, Ihering [MHNG, NHMB, examined].
Iridomyrmex dispertitus subsp. *micans* Forel. Forel 1911: 306 (Q). [mat. ref.

BRAZIL, São Paulo, Bahnhof Alto da Serra, Leuderwaldt, MHNG, examined].

Iridomyrmex humilis r. *platensis* Forel 1912: 46-47 (W, Q, M). Lectotype worker, by present designation [MHNG, examined], 13 worker, 5 queen, and 6 male paralectotypes, La Plata, Argentina, Bruch [MHNG, examined, lectotype series additionally contains a single misidentified *Paratrechina* worker]. **syn. nov.**

Iridomyrmex humilis r. *platensis* var. *transiens* Forel 1913: 46-47 (W, Q, M). Unavailable name (quadrinomen). [mat. ref. 15w, 5q, 9m, ARGENTINA, Buenos Aires, Atalaya, C. Bruch, MHNG, examined].

Iridomyrmex humilis Mayr st. *angulata* var. *pertaesta* Santschi 1916: 390 (W). Unavailable name (quadrinomen). [mat. ref. 15w, ARGENTINA, Buenos Aires, V. Steiger, NHMB, examined].

Iridomyrmex humilis st. *scotti* Santschi 1919: 52-53 (W, Q). Lectotype worker, by present designation [NHMB, examined], and 4 paralectotype workers, Cabana, Córdoba, Argentina, Scott [MACN, NHMB, examined]. **syn. nov.**

Linepithema dispertitum micans (Forel). Shattuck 1992a: 16. First combination in *Linepithema*.

Linepithema humile platense (Forel). Shattuck 1992a: 16. First combination in *Linepithema*.

Linepithema humile scotti (Santschi). Shattuck 1992a: 16. First combination in *Linepithema*.

Linepithema dispertitum micans (Forel). Shattuck 1994: 122.

Linepithema humile platense (Forel). Shattuck 1994: 124.

Linepithema humile scotti (Santschi). Shattuck 1994: 124.

Linepithema dispertitum micans (Forel). Bolton 1995: 247.

Linepithema humile platense (Forel). Bolton 1995: 247.

Linepithema humile scotti (Santschi). Bolton 1995: 247.

Species group: Humile

Lectotype worker measurements: HL 0.77, HW 0.68, MFC 0.19, SL 0.70, FL 0.64, LHT 0.69, ES 2.91, PW 0.48, CI 89, SI 102, CDI 27, OI 38.

Worker measurements: (n = 40) HL 0.50–0.84, HW 0.42–0.80, MFC 0.12–0.20, SL 0.46–0.79, FL 0.39–0.71, LHT 0.42–0.77, ES 1.03–3.38, PW 0.30–0.49, CI 84–95, SI 97-110, CDI 21–29, OI 20–40.

Worker diagnosis: Pubescence dense throughout mesopleuron, metapleuron, and gastric tergites; antennal scapes moderately long (SI 97–110), in repose usually exceeding posterior margin of head by a length greater than or equal to that of first funicular segment; first and second gastric tergites bearing numerous standing setae.

Worker description: Head in full face view somewhat longer than broad (CI 84–95), lateral margins convex, posterior margin straight to concave. Head normally reaches widest point at or posterior to level of compound eyes. Compound eyes moderate to large in size (OI 20–40), comprised of 60–105 ommatidia. Antennal scapes moderately long (SI 97–110), as long as or slightly shorter than head length. In full face view, scapes in repose usually exceed posterior margin of head by a length greater than or equal to that of first funicular segment. Frontal carinae narrowly to moderately spaced (CDI 21–29). Maxillary palps relatively short, approximately ½ HL or less, ultimate segment (segment 6) subequal in length or shorter than segment 2.

Pronotum and anterior mesonotum either forming a continuous curve, or curve interrupted by a slight medial mesonotal impression. Metanotal groove slightly to moderately impressed. Propodeum relatively high and inclined anteriorly, posterior propodeal face in profile broken at level of propodeal spiracle.

Petiolar scale relatively sharp and inclined anteriorly, in lateral view falling short of propodeal spiracle.

Cephalic dorsum (excluding clypeus) bearing 0–5 erect to suberect setae (mean = 1.3). Pronotum with 0–3 erect setae (mean = 1.1). Mesonotum without erect setae. Gastric tergite 1 (= abdominal tergite 3) bearing 2–10 erect to subdecumbent setae (mean = 5.1) including posterior row, tergite 2 bearing 2–16 erect setae (mean = 8.1), tergite 3 bearing 4–10 erect setae (mean = 5.9). Venter of metasoma with scattered erect setae.

Sculpture on head and mesosomal dorsum shagreened and slightly shining to moderately opaque. Pubescence dense on head, mesosoma and metasoma. Mesopleuron and metapleuron with dense pubescence.

Color variable, from testaceous to piceous, mesosoma sometimes lighter than head and gaster.

Queen measurements: (n = 8) HL 0.79–0.98, HW 0.75–0.95, SL 0.71–0.87, FL 0.70–0.91, LHT 0.79–1.00, EL 0.28–0.36, MML 1.55–2.38, WL 4.68–5.35, CI 95–100, SI 86–94, OI 33–37, WI 28–31, FI 38–48.

Queen description: Moderate to large species (MML > 1.5). Head slightly longer than broad to about as long as broad in full face view (CI 95–100), posterior margin concave. Eyes of moderate size (OI 33–37). Ocelli small to moderate in size. Antennal scapes of moderate length (SI 86–94), in full face view scapes in repose surpassing posterior margin by a length less than to slightly greater than length of first funicular segment.

Forewings moderately short relative to mesosomal length (WI 28–31). Forewings with Rs+M subequal to about twice length of M.f2. Legs of moderate length relative to mesosomal length (FI 38–48).

Dorsum of mesosoma and metasoma with numerous standing setae. Mesoscutum bearing 5–15 standing setae. Body color light brown to piceous. Antennal scapes, legs, and mandibles concolorous with or somewhat lighter than body.

Male measurements: (n = 6) HL 0.51–0.59, HW 0.49–0.58, SL 0.14–0.16, FL 0.52–0.65, LHT 0.45–0.54, EL 0.22–0.28, MML 0.98–1.24, WL 2.27–2.71, PH 0.19–0.27, CI 95–102, SI 26–30, OI 40–49, WI 21–24, FI 49–56.

Male diagnosis: Forewing with 1 submarginal cell; propodeum with strongly concave posterior face, overhanging petiole; mesosoma not greatly swollen (MML < 1.3); appendages not elongate (FI < 56); pubescence on head moderate to dense in region posterior to compound eye.

Male description: Head robust, about as long as broad in full face view (CI 95–102). Eyes moderate to large in size (OI 40–49), occupying much of anterolateral surface of head anterior of midline and separated from posterolateral clypeal margin by a length less than width of antennal scape. Ocelli small and in full frontal view set above adjoining posterolateral margins. Antennal scape moderately long (SI 26–30), about 3/4 length of 3rd antennal segment. Anterior margin of median clypeal lobe broadly convex. Mandibles variable, small to moderate in size, usually bearing a single apical tooth and 7–13 denticles along masticatory margin. Masticatory margin relatively narrow to relatively broad, longer than or subequal in length to inner margin. Inner margin and exterior lateral margin converging, parallel, or diverging.

Mesosoma moderately developed, slightly larger or subequal in bulk to metasoma. Mesoscutum enlarged, projecting forward in a convexity overhanging pronotum. Scutellum large, convex, nearly as tall as mesoscutum and projecting above level of propodeum. Propodeum well developed and overhanging petiolar node, posterior propodeal face strongly concave. Forewings of moderate length relative to mesosomal length (WI 21–24) and bearing a single submarginal cell. Wing color clear to slightly smoky with light to dark brown veins and stigma. Legs moderately short relative to mesosoma length (FI 49–56).

Petiolar scale sharp and taller than node length. Ventral process well developed and pointing posteriorly. Gaster oval in dorsal view, nearly twice as long as broad. Gonostylus produced as bluntly rounded pilose lobes. Volsella with cuspis present, digitus short and downturned distally.

Dorsal surfaces of body largely devoid of erect setae, mesoscutum lacking standing setae, posterior abdominal tergites with a few fine, short setae. Venter of gaster with scattered setae. Pubescence dense on body and appendages, becoming sparse only on medial propodeal dorsum and lateral faces of pronotum.

Head, mesosoma and metasoma light to dark brown. Legs, mandibles, and antennae often lighter in color.

Distribution: Central Argentina to eastern Brazil.

Biology: Linepithema micans has been recorded from sea level to 2300 meters elevation. Where habitat information has been noted, five records are from pasture or grassland, two from high elevation semideciduous forest, two from wet montane forest, and two from second growth riparian forest. Seven nests were collected from under stones, two from rotting wood, and three from sandy soil. *Linepithema micans* has also been taken at baits, in pitfall traps, and in sifted litter. Alates have been recorded in the nests in Argentina from August to January, and in Brazil year round. Orr and Seike (1998) report that this species (as. "*L. humile*", voucher specimens at UCDC examined) is attacked by *Pseudacteon pusillum* phorid flies at several locations in Brazil, and that these attacks alter the outcome of competitive interactions with other ant species at baits.

Leuderwaldt (1926: 286) reported the following observations on this species (as "*Iridomyrmex dispertitus* For. subsp. *micans* For.", and translated from Portuguese):

"Alates in the nest: April. Colonies in natural habitats under rocks; small colonies, containing around 50 workers. A nest was found on the trunk of a *Cyathea schanschin* Mart. between the leaf stalks, close to a nest of *L. leucomelas*. In nest #10362 many staphylinids *Aleochara leuderwaldti* Bernh. were found; in another #11,942, *Atheta tuberculicauda* Bernh. with 5 from nest #2,650 under a rock, in Campo do Itatiaya (Rio)." Given the polydomy of other *Linepithema* species, I suspect Leuderwaldt's small colony size estimate refers not to colony size but perhaps to nest size within a larger colony.

Similar species: Workers of *L. gallardoi* have mesopleural pubescence fading to sparse anteroventrally and often have slightly shorter legs and antennal scapes (Figs. 83–85). *Linepithema humile* workers lack standing setae on gastric tergites 1–2 and usually have larger eyes (Fig. 86). *Linepithema oblongum* and *L. anathema* workers are more gracile with longer antennal scapes (SI > 115, Figs. 83–85). *Linepithema neotropicum* workers have much longer maxillary palps, a less pubescent mesopleura, and a lower, more rounded propodeum (Fig. 29).

Discussion: *Linepithema micans* is morphologically a rather generalized species that appears to form the paraphyletic core of the *humile* complex from which the more distinctive *L. humile*, *L. oblongum*, and *L. gallardoi* arose (Wild, molecular data). The boundaries of *L. micans* remain problematic, and there are possibly several cryptic species within the current conception of this taxon. This possibility is supported by broad elevational and habitat differences among populations of this species.

Forel's type workers of *L. micans* from São Paulo are large and medium reddish brown in color, with a slight medial mesonotal impression (Fig. 37) and relatively short pubescence. Specimens closely similar to the *micans* type, including many of the ants observed in Orr and Seike's (1998) ant-phorid study, are among the most commonly encountered *Linepithema* in São Paulo state. Judging from specimens at MZSP, Borgmeier was apparently misled by the local abundance of *L. micans* in southern Brazil and mistook *L. micans* for *L. humile*, consequently failing to recognize true *L. humile* and unnecessarily naming a new species, *L. riograndense*, for a population of Argentine ants from Rio Grande do Sul (synonymized by Wild [2004]).

Populations outside the Atlantic forest region are more divergent, and some of them may later merit elevation to specific status. Towards the south into Paraguay and Argentina workers of most populations become dark brown in color, often showing a slightly more robust build and lacking a mesonotal impression, as in Santschi's *L. scotti* types from Córdoba and Forel's *L. platense* types (Figs. 39–40) from La Plata. Workers from montane populations near Tucumán, Argentina and some from Minas Gerais are lighter in color with large eyes and unusually broad heads.

Material examined: ARGENTINA. Buenos Aires: Atalaya [MHNG]; Buenos Aires [MHNG, NHMB]; Costanera Sur, 34°37'S 58°21'W, 62m [ALWC, BMNH, IFML, MCZC, USNM]; La Plata [MHNG]; Olavarría [MHNG]; Plaza La Valle, 34°38'S 58°21'W [UCDC]; Rio Santiago [MACN, MHNG]; Sierra de la Ventana [NHMB]; Tandil [NHMB]; Buenos Aires (s.loc.) [MACN]. Córdoba: Cabana [MACN, NHMB]. Entre Rios: P. N. El Palmar, 31°53'S 58°13'W [ALWC, AVSC]. Jujuy: Calilegua [IFML]; Jujuy (s.loc.) [IFML]. Misiones: El Dorado [IFML]; Esperanza [IFML]; Iguazú [ALWC, IFML]; L. Alem. [IFML]; M. Belgrano [IFML]; Misiones (s.loc.) [MACN]. Tucumán: El Mollar, 26°56'S 65°41'W, 1800m [IFML, PSWC]; Parque Menhires, 26°56'S 65°41'W, 1700m [UCDC]; Tafi del Valle, 26°54'S 65°41'W, 1900m [ALWC, BMNH, IFML, LACM, MCZC, MZSP]; Tucumán, 26°56'S 65°41'W, 1800m [USNM]; Tucumán (s.loc.) [IFML]. Argentina (s.loc.) [IFML].

BRAZIL. Bahia: Encruzilhada da BA, 980m [MZSP]. Distrito Federal, Res. Biol. Águas Emendadas [MZSP]. Espírito Santo: Santa Teresa, 670m [MCZC, MZSP]. Goiás: 7 km NW Alto Paraiso, Morro das Cobras [MZSP]. Minas Gerais: 2 km S Monte Verde, 22°54'S 46°03'W, 1900m [MCZC, PSWC, UCDC]; Monte Verde, 22°52'S 46°03'W [UCDC]; Ouro Preto [USNM]; Serra Caraça, 1380m [MCZC, MZSP]. Mato Grosso do Sul: 8 km SE Ponta Pora [WPMC]. Paraná: Caiobá [MZSP]; Ponta Grossa, Villa Velha [MZSP]; Rio Negro [MZSP, NHMB]; Três Pinheiros, 26°42'S 51°34'W, 1200m [MZSP]. Rio de Janeiro: Floresta da Tijuca [MZSP]; Itatiaia [MZSP, NHMB, USNM]; Mendes [MZSP]; Petrópolis [MZSP]; Petrópolis, Tr. da Fatima [MZSP]. Rio Grande do Sul: Bom Jesus [MZSP]; Caxias do Sul-V.N. Conceição [MZSP]; Cotegipe [MZSP]; Tainhas [MZSP]; Rio Grande do Sul (s.loc.) [MHNG, MCSN]. Santa Catarina: Blumenau [NHMW]; Chapecó [MZSP]; Florianopolis, Praia da Joaquina [MZSP]; Forquilhinha, Mun. Criciuma [MZSP]; N. Teutônia, 300-500m [MZSP]; P. Borman [MZSP]; Seara [MZSP]; Serra Geral [MZSP]; Xaxim [MZSP]; Santa Catarina (s. loc.) [BMNH, NHMW]. São Paulo: Alto da Serra [MHNG, MZSP]; Barueri [MZSP]; BR 116 km 40 S. Paulo-Curutiba [MZSP]; Campos do Jordão [MZSP, UCDC]; BR SP Caraguatatuba, 680m [MZSP]; Cunha, P.E. Serra do Mar., Nucleo Cunha Indaia, 23°15'S 45°00'W [MZSP]; Estr. V. Santos, Meio da Serra [MZSP]; Guarerama [MZSP]; Paranapiacaba [MZSP]; Parelheiros [MZSP]; Repr. Jurubatuba, V. Estr. Santos [MZSP]; Salesópolis, Est. Biol. Boraceia, 850m [MCZC, MZSP]; Santana de Parnaiba [MZSP]; São Paulo, Agua Funda [MZSP]; Serra Cubatão, V. Estr. Santos [MZSP]; Serra do Japi, 23°16'S 47°00'W [UCDC]; São Paulo (s.loc.) [MHNG, NHMB, NHMW]. PARAGUAY. Central: Capiatá [ALWC]. URUGUAY. Uruguay (s.loc.) [NHMB].

Linepithema neotropicum Wild, **sp. nov.**

(worker mesosoma Figs. 6, 29; worker head Fig. 30; male head Fig. 65; distribution Fig. 104)

Species group: Neotropicum

Holotype worker. PARAGUAY. Canindeyú: Reserva Natural del Bosque Mbaracayú, Jejuimi, 24°08'S 055°32'W, 12.xi.2002. A.L.Wild acc. no. AW1718. [1 worker, INBP].

Paratypes. Same collection data as holotype, A. L. Wild acc. nos. AW1718 [1022 workers, 30 males, 29 queens, ALWC, BMNH, CASC, JTLC, IFML, INBP, MCZC, MHNG, MZSP, NHMB, NHMW, PSWC, QCAZ, UCDC, USNM, WPMC].

Holotype worker measurements: HL 0.66, HW 0.60, MFC 0.16, SL 0.60, FL 0.52, LHT 0.55, ES 1.97, PW 0.38, CI 91, SI 101, CDI 27, OI 29.

Worker measurements: (n = 31) HL 0.54–0.70, HW 0.47–0.66, MFC 0.12–0.18, SL 0.50–0.64, FL 0.43–0.56, LHT 0.47–0.59, ES 1.24–2.37, PW 0.31–0.44, CI 87–96, SI 97-110, CDI 24–29, OI 23–34.

Worker diagnosis: Maxillary palps relatively long (> ½ HL; segment 6 longer than segment 2); propodeum low and rounded; metapleural bulla with at least some appressed pubescence.

Worker description: Head in full face view slightly longer than broad (CI 87–96), lateral margins convex, posterior margin concave. Compound eyes of moderate size (OI 23–34), comprised of 50–70 facets. Antennal scapes of moderate length (SI 97–110), shorter than head length. In full face view, scapes in repose surpass posterior margin of head by an amount subequal to the length of the first funicular segment. Frontal carinae narrowly to moderately spaced (CDI 24–29). Maxillary palps relatively long, greater than ½ HL, ultimate segment (segment six) longer than segment 2.

Mesosoma in lateral view with pronotum and mesonotum forming a single continuous convexity, mesonotal profile straight to very slightly impressed mesally. Metanotal groove slightly impressed to not at all impressed, propodeum low and rounded, dorsal and posterior faces rounding evenly into each other without a distinct angle.

Petiolar scale sharp and inclined anteriorly, in lateral view falling short of the propodeal spiracle.

Cephalic dorsum (excluding clypeus) lacking erect setae. Pronotum with one pair of erect setae. Mesonotum without erect setae. Gastric tergite 1 bearing 1–3 erect setae (mean = 1.7), tergite 2 with 1–4 erect setae, tergite 3 with 2–5 erect setae. Venter of metasoma with scattered erect setae.

Surface of head and mesosomal dorsum shagreened and moderately shining. Pubescence dense on head and mesosomal dorsum. Mesopleura with pubescence fading from moderate posterodorsally to sparse or absent anteroventrally. Metapleura with moderate to sparse pubescence thinner than on propodeal dorsum but never completely absent. Gastric tergites 1–4 with dense to moderate pubescence, surface moderately shining.

Body color light brown to dark brown, area on head surrounding mandibular insertions often lighter. Trochanters and tarsi pale whitish-brown.

Queen measurements: (n = 5) HL 0.78–0.82, HW 0.70–0.73, SL 0.70–0.73, FL 0.69–0.73, LHT 0.75–0.80, EL 0.28, MML 1.59–1.72, WL 4.85–5.33, CI 89–91, SI 97–103, OI 34–36, WI 30–31, FI 43–45.

Queen description: Moderately small species (MML < 1.8). Head narrow in full face view (CI 89–91), posterior margin straight. Eyes moderate to small (OI 34–36). Ocelli small. Antennal scapes moderate to long (SI 94–101), in full face view scapes in repose surpassing posterior margin by a length approximately that of first funicular segment.

Forewings moderately long relative to mesosomal length (WI 30–31). Forewings with Rs+M more than two times longer than M.f2. Legs of moderate length relative to mesosomal length (FI 43–45).

Dorsum of mesosoma and metasoma with scattered suberect standing setae. Mesoscutum bearing 0–6 standing setae. Body color medium brown. Trochanters and tarsi lighter.

Male measurements: (n = 5) HL 0.46–0.51, HW 0.45–0.50, SL 0.12–0.13, FL 0.49–0.54, LHT 0.44–0.48, EL 0.20–0.23, MML 0.91–1.02, WL 2.32–2.54, PH 0.16–0.20, CI 95–101, SI 24–27, OI 40–47, WI 25–26, FI 52–56.

Male diagnosis: Forewing with 1 submarginal cell; petiolar node with dorsal scale taller than long in lateral view; posterior face of propodeum straight to slightly concave; sculpture of head not well developed and surface slightly shining through pubescence; mandibles with apical tooth unusually elongate; size moderate (MML 0.91–1.02).

Male description: Head about as broad as long in full face view (CI 95–101). Eyes of moderate size (OI 40–47), occupying much of anterolateral surface of head and separated from posterolateral clypeal margin by a length less than width of antennal scape. Ocelli of moderate size and in full frontal view set above adjoining posterolateral margins. Antennal scape short (SI 24–27), about 2/3 length of 3rd antennal segment. Anterior clypeal margin convex medially. Mandibles moderately sized and somewhat elongate, masticatory margin broad, longer than inner margin, apical tooth enlarged as a sharp spine and followed by a series of 8–18 denticles. Inner margin and exterior lateral margin parallel to slightly diverging.

Mesosoma moderately developed and subequal in length to metasoma. Mesoscutum not greatly enlarged, projecting slightly forward over pronotum. Propodeum in lateral view not strongly overhanging petiole, posterior face slightly convex to slightly concave. Forewings of moderate length (WI 25–26) and bearing one submarginal cell.

Wings transparent, with pale brown wing veins and stigma. Legs of moderate length relative to mesosoma (FI 52–56).

Petiolar node bearing an erect scale, node height taller than node length. Venter of node bearing a convex downward-pointing process. Gaster ovoid in dorsal view, about 2 times as long as broad. Gonostylus produced as a pointed, triangular pilose lobe. Volsella with cuspis present, digitus short and downturned distally.

Dorsal surfaces of body with erect setae sparse, mesoscutum with 0–4 fine erect to suberect setae. Venter of gaster with scattered setae. Pubescence dense on body and appendages, becoming sparse only on medial propodeal dorsum. Sculpture on head and mesoscutum not well developed, surface shining through pubescence.

Head dark brown. Mesosoma and metasoma pale brown to medium brown. Mandibles, antennae, and legs pale whitish-yellow to light brown, much lighter than body.

Distribution: Costa Rica south to Paraguay and southeastern Brazil.

Biology: This species inhabits a considerable range of forest habitats from sea level to over 2000 meters. Most records are from lowland tropical or subtropical humid forests, but several are from forest edge, 2nd growth, or logged primary forest. Specimens collected in the central Brazilian state of Tocantins are from cerrado or cerradão habitats. Five records are from park or garden habitats, and there are single collections of *L. neotropicum* from coffee, soy, cacao, banana, and pineapple fields. Six nest records are from soil, two from under stones, one in a rotting log, and one in soil under a rotting log. Soil nests often have several inconspicuous entrances not much larger than the width of a worker ant, with small piles of excavated earth around them (Wild, pers obs.).

Linepithema neotropicum, like most species, appears to be a trophic generalist. Ants have been observed recruiting to tuna, sardine, and honey baits in Colombia, Brazil, and Paraguay. One collection records root-feeding pseudococcids in a soil nest and another documents aphid-tending on vegetation above ground. Middens from a nest in Ecuador contained fragments of dead arthropods including nasute termites and *Trachymyrmex* ants.

The type series was collected by the author as an entire colony from the shaded north side of a laboratory building in Paraguay at the edge of a humid subtropical semideciduous forest. The colony had scaled the side of the building during the day as if to avoid a subterranean army ant predator (Gotwald 1995, pg 229), although no army ants were seen. The full colony consisted of 1022 worker ants, one dealate queen, 28 alate queens, 30 males, and 82 additional male pupae.

Alate males and females were observed flying to a fluorescent light in mid November at the Mbaracayú Reserve in Paraguay, between 21:00 and 23:00hr on a warm evening after two days of rain. Alate females have been taken in nests in November in Paraguay, alate males have been recorded from nests in Ecuador in December, and alate males and females have been collected in pan traps and in low vegetation in Rondônia, Brazil in November and December.

Similar species: Workers of the closely-related *L. cerradense* have a more flattened mesosomal profile (Fig. 27), a narrower head, and shorter antennal scapes (Fig. 81). Workers of *L. dispertitum*, from Central America, normally have a more deeply impressed metanotal groove (Fig. 21). Humile-group species have much shorter maxillary palps. Males of *L. cerradense* are similar in structure to *L. neotropicum* males but tend to be lighter in color and smaller in size (MML < 0.80, Fig. 93).

Discussion: This is a widespread and variable species that has been misidentified in various collections as *L. iniquum* (MHNG), *L. humile* (NHMB), *L. scotti* (MZSP, NHMB), *L. leucomelas* (NHMB), and *L. dispertitum* (MCSN). Most worker specimens are dark brown in color, but in northeastern Brazil there are some color and pilosity intergrades with *L. cerradense*, perhaps not surprising considering that *L. neotropicum* is possibly paraphyletic with respect to *L. cerradense* (Wild, molecular data). Some high elevation collections from Colombia and the records from Trinidad and Guyana are also lighter in color. Eye size varies somewhat. Some Costa Rican specimens and specimens from the Atlantic forest region of Brazil and Paraguay have the largest eyes (ES around 2.0, 60–70 ommatidia), while specimens from northwestern South America have the smallest eyes (ES around 1.5, 50–60 ommatidia). Costa Rican specimens are more variable in pilosity, sometimes bearing more than 2 erect setae on the pronotum. The extent of mesopleural pubescence varies over a north-south cline, with specimens from the northern part of the range having more extensive pubescence.

Etymology: The name refers to this ant's wide distribution in the neotropics.

Material examined: BOLIVIA. Cochabamba: 67.5 km E. Villa Tunari Valle Sajta [ALWC, UCDC, WPMC]. La Paz: Cañamina [USNM]; Coroico, Chilumani, Yungas 1600m [MCSN]. Santa Cruz: Santa Cruz, 17°46'S 63°11'W 420m [PSWC]. BRAZIL. Amazonas: 61 km N Manaus on Caracaraí Rd [MCZC]; Manaus, Reserva Ducke [MCZC]. Bahia: Confl. R. Mucugeo & R. Cumbuca [MZSP]; Encruzilhada, 960m [MZSP]; Itabuna [MZSP]; Itamaruju [MZSP]; Salvador [MZSP]. Espírito Santo: Linhares [MZSP]; Pedro Canário [MZSP]; Vila Velha [MZSP]. Goiás: Alto Paraiso.

Faz Boa Espera [MZSP], Anápolis [MZSP]. Maranhão: Balsas, Gerais de Balsas, 08°37.450'S 046°44.500'W [MZSP]. Mato Grosso: Chapada [MZSP]; Colonia Vicentina Dourados [MZSP]; Sto. Antonio de Leverger, Aguas Quentes [MZSP]. Minas Gerais: Belo Horizonte [USNM]; S. Caraça (Engenho) 800m [MZSP]; Sabará [USNM]; Varginha [MZSP]; Viçosa [MZSP]. Paraná: Foz do Iguaçu [USNM]. Pernambuco: Caruaru 900m [MZSP]. Rio de Janeiro: Conv. S. Antonio [MZSP]; Floresta de Tijuca [MZSP]; Friburgo [MZSP]; Ilha Grande, 23°10'S 44°10'W [ALWC]; Itatiaia [MZSP]; Marambaia [MZSP]; Maringá, Cachoeira Sta. Clara [MZSP]; Mendes [MZSP, NHMB, USNM]; Petrópolis [MZSP]; Reserva Biológica União, 22°27'S 42°02'W [UCDC]; Rio de Janeiro, DF Grajaú [MZSP]; Silva Jardim [MZSP]; Rio de Janeiro (s. loc.) [NHMB]. Rondônia: Faz. Rancho Grande, 10°18'S 62°53'W [ALWC, UCDC]. São Paulo: 30k NE São Paulo, Sitio S. Francisco [MZSP]; Amparo [MZSP]; Anhembi, Faz. B. Rico [MZSP]; Cajurú, Faz. Santa Carlota, 21°17'S 047°18'W [MZSP]; Mata de Santa Genebra, 22°49'S 47°6'W [UCDC]; Lençois [MZSP]; Pauba, S. Sebastião [MZSP]; Pedra Azul, 800m [MZSP]; Piracicaba [MZSP]; Ribeirão Pires [USNM]; S. Sebastião [MZSP]; S. Sebastião, B. S. Francisco [MZSP]; São Paulo [MZSP]; São Paulo, Butantan [MZSP]; São Vicente [USNM]; Ubatuba [MZSP]. Tocantins: Babaçulandia, 07°01.050'S 47°52.233'W [MZSP]; Porto Nac'l Faz Alto Paraiso, 10°43.533'S 48°23.083'W [MZSP]; Porto Nac'l Faz Sucupira, 10°39.733'S 48°44.483'W [MZSP]. COLOMBIA. Huilá: 3 km E Rivera [WPMC]; 4k NE Rivera [WPMC]. La Guajira: San Antonio [MHNG]. Meta: 8 mi W. Villavicencio, 1050m [CASC]. Tolima: Transecto Parque los Nevados, E slope 500m [PSWC]; Transecto Parque los Nevados, E slope 1150m [PSWC]; Transecto Parque los Nevados, E slope 1670m [PSWC]; Transecto Parque los Nevados, E slope 2020m [PSWC]; Transecto Parque los Nevados, E slope 2180m [PSWC]. Valle: 19 mi. S. Cartago, 960m [CASC]; Buga [WPMC]; Cali, Universidad del Valle Campus [ALWC]. COSTA RICA. Heredia: Est. Biol. La Selva, 10°26'N 84°01'W, 150m [JTLC]; San Pedro de Barva [LACM]; Sto. Domingo, 09°59'N 84°05'W, 1100m [JTLC, LACM]; vic. Heredia, 1000-1500m [JTLC]. Puntarenas: 6 km S Monteverde, 10°15'N 084°49'W, 800m [JTLC]. San José: San José, 09°56'N 084°5'W, 1100m [JTLC]. ECUADOR. Napo: 1 km SW Archidona, W side of Rio Misahualli, 00°55'S 77°49'W, 550m [ALWC]; 3 km NNE Archidona, Road Baeza-Archidona, 00°51'S 77°48'W, 650m [ALWC]; 4 km SW Archidona, W side of Rio Misahualli, 00°55'S 77°50'W, 550m [ALWC]; 6 km SE Archidona, Monteverde Ecological Reserve, 00°56'S 77°45'W, 620m [ALWC, BMNH, MCZC, MZSP, QCAZ, USNM]; Carlos J. Arosemena Tola, banks of Rio Anzu, 01°12'S 77°53'W 500m [ALWC]. GUYANA. Cuyuni-Mazaruni: Bartica [MCZC]. PARAGUAY. Alto Paraná: Ciudad del Este [INBP]. Canindeyú: Res. Nat. del Bosque Mbaracayú, Jejuimí, 24°08'S 55°32'W, 170m [ALWC, BMNH, CASC, JTLC, IFML, INBP, MCZC, MHNG, MZSP, NHMB, PSWC, QCAZ, UCDC, USNM, WPMC]; Res. Nat. del Bosque Mbaracayú, Aguara Ñu, 24°00'S 55°19'W, 500m [ALWC]. PERU. Cajamarca: Rio Charape [MCZC]. Junín: Chanchamayo, Anashirone R. [MZSP]. San Martín: Convento, 26 km NNE

Tarapoto, 06°16'S 76°18'W [PSWC]. TRINIDAD AND TOBAGO. St. George: Arima Valley [MCZC]. Trinidad (s.loc.) [LACM]. VENEZUELA. Aragua: Rancho Grande, 1100m [WPMC]. Lara: Guarico [UCDC]. Trujillo: 15 km ESE Boconó, 09°11'N 70°09'W, 1160m [MCZC].

Linepithema oblongum (Santschi)

(worker mesosoma Fig. 41; worker head Fig. 42; male head Fig. 73; distribution Fig. 101)

> *Iridomyrmex humilis* var. *oblongus* Santschi 1929: 306 (W). Holotype worker, Purmamarca, Jujuy, Argentina, J. Witte [NHMB, examined].
>
> *Iridomyrmex humilis* var. *oblungus* Santschi. Menozzi 1935: 321. Misspelling of *oblongus*. Possible misidentification of *L. humile* (Mayr).
>
> "*Iridomyrmex*" *oblonga* Santschi. Snelling and Hunt 1975: 90. Raised to species.
>
> *Linepithema oblongum* (Santschi). Shattuck 1992a: 16. First combination in *Linepithema*.
>
> *Linepithema oblongum* (Santschi). Shattuck 1994: 126.
>
> *Linepithema oblongum* (Santschi). Bolton 1995: 247.

Species group: Humile

Worker measurements: (n = 13) HL 0.61–0.75, HW 0.50–0.65, MFC 0.14–0.18, SL 0.65–0.83, FL 0.54–0.72, LHT 0.62–0.79, PW 0.35–0.44, ES 1.7–2.8, SI 120–139, CI 81–88, CDI 25–28, OI 28–38.

Worker diagnosis: Antennal scapes very long (SI > 120); mesopleura and metapleural bulla with dense pubescence; gastric pubescence usually fading to sparse on tergites 2–4.

Worker description: Head in full face view slender, longer than broad (CI 81–88), somewhat oval in shape, reaching widest point at or just posterior to compound eyes. Lateral margins broadly convex, posterior margin straight to weakly concave. Anterior clypeal margin with a broad, relatively shallow medial excision. Compound eyes of moderate size (ES 1.7–2.8), comprising 60–85 ommatidia. Antennal scapes long (SI 120–139), longer than HL and easily surpassing posterior margin of head in full face view. Frontal carinae moderately spaced (CDI 25–28). Maxillary palps relatively short, segments 4, 5 and 6 each noticeably shorter than segment 2.

Pronotum and mesonotum forming a single convexity. Posterior 2/3 of pronotal dorsum in lateral view straight or only slightly convex. Mesonotal dorsum nearly straight, not angular or strongly impressed, frequently with a slight medial impression. Metanotal groove slightly to moderately impressed. Propodeum in lateral view quadrate-forward.

Petiolar scale sharp and inclined anteriorly, in lateral view falling short of the propodeal spiracle.

Dorsum of head and mesosoma devoid of erect setae. Gastric tergites 1–2 (= abdominal tergites 3 and 4) usually without erect setae, occasionally with 1–2 erect setae (mean = 0.1). Gastric tergite 3 with 2–5 erect setae (mean = 3.5). Venter of metasoma with scattered erect setae.

Head, mesosoma, and appendages covered in dense pubescence. Mesosoma and metapleural bulla with dense pubescence. Pubescence on gaster variable, even within a nest series, nearly always dense on gastric tergite 1 (= abd. tergite 3), usually fading to sparse on tergites 2–4, surface relatively shining.

Body weakly to strongly bicolored, gaster always darker than head and mesosoma. Color varies from head and mesosoma testaceous and gaster brown to head and mesosoma brown and gaster dark brown.

Queen measurements: (n = 2) HL 0.83–0.85, HW 0.75–0.78, SL 0.84–0.86, FL 0.81–0.82, LHT 0.91, EL 0.27–0.28, MML 1.55–1.58, WL 4.89–4.91, CI 90–91, SI 110–112, OI 33, WI 31–32, FI 52.

Queen description: Moderately small species (MML 1.55–1.58). Head narrow in full face view (CI 90–91), posterior margin straight to slightly concave. Eyes moderately small (OI 33). Ocelli moderately small. Antennal scapes long (SI 110–112), in full face view scapes in repose surpassing posterior margin by a length greater than length of first funicular segment.

Forewings of moderate length relative to mesosomal length (WI 31–32). Forewings with Rs+M at least two times longer than M.f2. Legs long relative to mesosomal length (FI 52).

Dorsum of mesosoma and metasoma with scattered standing setae. Mesoscutum bearing 0–4 suberect setae. Body color medium brown to dark brown, gaster often somewhat darker than mesosoma. Antennal scapes, legs, and mandibles concolorous with body.

Male measurements: (n = 4) HL 0.50–0.54, HW 0.47–0.49, SL 0.13–0.15, FL 0.55–0.57, LHT 0.51–0.52, EL 0.22–0.23, MML 0.98–1.01, WL 2.29–2.59, PH 0.23–0.24, CI 91–95, SI 24–29, OI 40–44, WI 23–26, FI 56–57.

Male diagnosis: Forewing with 1 submarginal cell; propodeum with strongly concave posterior face, overhanging petiole; mesosoma not greatly swollen (MML < 1.3); head clearly longer than broad in full face view (CI 91–95); compound eyes small (OI 40–44); appendages somewhat elongate (FI > 56).

Male description: Head longer than broad in full face view (CI 91–95). Eyes small (OI 40–44), occupying much of anterolateral surface of head anterior of midline and separated from posterolateral clypeal margin by a length less than width of antennal scape. Ocelli small and in full frontal view set above adjoining posterolateral margins. Antennal scape of moderately length (SI 24–29), about 2/3 length of 3rd antennal segment. Anterior clypeal margin straight to broadly convex. Mandibles variable, small to moderate in size, usually bearing a single apical tooth, sometimes a strong subapical tooth, and 8–12 denticles along masticatory margin. Masticatory margin relatively narrow to relatively broad, longer than or subequal in length to inner margin. Inner margin and exterior lateral margin converging, parallel, or diverging.

Mesosoma well developed, larger in bulk than metasoma. Mesoscutum enlarged, projecting forward in a convexity overhanging pronotum. Scutellum large, convex, nearly as tall as mesoscutum and projecting well above level of propodeum. Propodeum well developed and overhanging petiolar node, posterior propodeal face strongly concave. Forewings of moderate length relative to mesosomal length (WI 23–26) and bearing a single submarginal cell. Wing color smoky with dark brown veins and stigma. Legs of moderate length relative to mesosoma length (FI 56–57).

Petiolar scale taller than node length. Ventral process well developed and pointing posteriorly. Gaster oval in dorsal view, nearly twice as long as broad. Gonostylus produced as bluntly rounded pilose lobes. Volsella with cuspis present, digitus short and downturned distally.

Dorsal surfaces of body largely devoid of erect setae, mesoscutum lacking standing setae, posterior abdominal tergites with a few fine, short setae. Venter of gaster with scattered setae. Pubescence dense on body and appendages, becoming sparse only on medial propodeal dorsum.

Head, mesosoma and metasoma medium to dark brown. Legs, mandibles, and antennae medium brown, trochanters lighter in color.

Distribution: Andean regions of Bolivia and northern Argentina.

Biology: Little is known about this species. All collection records are montane, between 1300 and 3800 meters elevation, and from open habitats such as roadsides, pastures, and alpine grasslands with one record from an urban park/garden in La Paz, Bolivia. I observed several nests of this species along a roadside in grazed alpine grassland at 2600 meters elevation near Tafí del Valle, Argentina, in October 2002. Three colonies were found under stones and one in soil marked by two small entrances ringed with excavated earth. Two nests contained alate males and females. Foragers were active during the day and at night, sometimes forming dilute trails.

Similar species: Workers of the Argentine ant *L. humile*, the sister species, are less gracile, usually bearing somewhat shorter scapes (SI < 127, Figs. 83, 85), lack standing setae on gastric tergites 1–2, and always have dense pubescence on all gastric tergites. Workers of *L. anathema* have a more prominent propodeum (Fig. 33), and have dense pubescence on all gastric tergites. Workers of the similarly elongate species *L. iniquum* and *L. leucomelas* have greater than 4 standing setae on the cephalic dorsum.

Discussion: Molecular evidence (Wild, unpublished) strongly supports a sister group relationship between *L. oblongum* and the morphologically similar pest species *L. humile*. This finding should elevate the importance of *L. oblongum* for comparative studies on the biology of the Argentine ant.

Specimens from Bolivia are usually somewhat lighter in color than those from Argentina.

Wilhelm Goetsch collected a *Linepithema* species from Copiapó, Chile, that was identified by Menozzi (1935: 321) as *L. oblongum*. This Chilean record is plausible, but given *L. oblongum*'s morphological similarity to the invasive *L. humile* which is abundant in Chile, the identity cannot be verified. Snelling and Hunt (1975) reached a similar conclusion. Goetsch himself identified his Copiapó ants as conspecific with the invasive Argentine ant, but his line drawings comparing the Copiapó ant with European *L. humile* show the slightly smaller eyes characteristic of *L. oblongum* (Goetsch 1957: 43). Further collections in northern Chile should clarify the distribution of this ant.

Material examined: ARGENTINA. Jujuy: Lagunas Yala [IFML]; Purmamarca [NHMB]; Jujuy (s.loc.) [IFML]. Tucumán: Aconquija [IFML]; Infiernillos [IFML]; Tafí del Valle, Carapunco, 26°47'S 065°43'W, 1600m [IFML, UCDC]; 7 km N Tafí del Valle, 26°47'S 065°43'W, 2700m [ALWC, BMNH, CASC, IFML, LACM, MCZC, MHNG, MZSP, QCAZ, USNM]; Quebrada de la Mesada, 26°22'S 065°32'W, 1300m [UCDC]. BOLIVIA. Cochabamba (s.loc.) [IFML]; Bolivia, "Cochabamba, La

Paz" [IFML]. La Paz: La Paz [MCZC, PSWC]. Oruro: Playa Verde, c. 20k S Oruro [BMNH].

Linepithema piliferum (Mayr)

(worker mesosoma Fig. 17; worker head Fig. 18; male head Fig. 61; male volsella Fig. 62; distribution Fig. 105)

> *Hypoclinea pilifera* Mayr 1870a: 393 (W, Q). Holotype worker, Colombia (as "Neu Granada") [NHMW, examined].
> *Hypoclinea (Iridomyrmex) pilifera* Mayr. Mayr 1870b: 954.
> *Iridomyrmex pilifer* (Mayr). Dalla Torre 1893: 169. First combination in *Iridomyrmex*. [unjust. emend.]
> *Linepithema piliferum* (Mayr). Shattuck 1992a: 16. First combination in *Linepithema*.
> *Linepithema piliferum* (Mayr). Shattuck 1994: 127.
> *Linepithema piliferum* (Mayr). Bolton 1995: 247.

Species group: Fuscum

Worker measurements: (n = 12) HL 0.65–0.83, HW 0.56–0.79, MFC 0.17–0.23, SL 0.61–0.79, FL 0.52–0.71, LHT 0.54–0.79, ES 1.42–2.93, PW 0.39–0.54, CI 86–95, SI 99–120, CDI 26–31, OI 21–36.

Worker diagnosis: A large species (HW > 0.55); antennal scapes long (SI 99–120), in repose exceeding posterior margin of head by a length greater than or equal to length of first funicular segment; cephalic dorsum bearing at least 2, and often more, erect to subdecumbent setae; mesonotum without strong medial impression (sometimes weakly impressed); mesopleura and metapleura lacking pubescence and strongly shining.

Worker description: Head in full face view somewhat longer than broad (CI 86–95), lateral margins evenly convex, posterior margin straight to slightly concave. Head normally reaches widest point posterior of compound eyes. Compound eyes moderate to large in size (OI 21–36), comprised of 50–75 ommatidia. Antennal scapes relatively long (SI 99–120), normally slightly shorter than head length, rarely slightly longer than head length. In full face view, scapes in repose exceed posterior margin of head by a length greater than or equal to length of first funicular segment. Frontal carinae moderately spaced (CDI 26–31). Maxillary palps of moderate length, approximately ½ HL, ultimate segment (segment 6) variable, somewhat shorter to somewhat longer than segment 2.

Pronotum and anterior mesonotum forming a continuous curve. Mesonotal dorsum relatively straight to moderately angular, often with a slight medial impression. Metanotal groove not impressed to moderately impressed. Dorsal propodeal face straight to slightly convex in lateral view. Posterior propodeal face convex.

Petiolar scale relatively sharp and inclined anteriorly, in dorsal view relatively narrow, in lateral view falling short of propodeal spiracle.

Cephalic dorsum (excluding clypeus) with numerous standing setae, 0–6 (mean = 3.8) erect to subdecumbent setae near antennal insertions and 0–3 (mean = 1.6) erect to subdecumbent setae near vertex. Pronotum with 0–4 erect to subdecumbent setae (mean = 2.3). Mesonotum without standing setae, or rarely with 1 small erect seta. Propodeal dorsum usually without standing setae, rarely with several erect setae. Gastric tergite 1 (= abdominal tergite 3) bearing 0–4 erect to suberect setae (mean = 2.3) mesally, exclusive of a row of 5–10 subdecumbent setae along posterior margin of tergite, tergite 2 bearing 2–6 erect setae (mean = 4.3) exclusive of posterior row, tergite 3 bearing 3–7 erect setae (mean = 4.8) exclusive of posterior row. Venter of metasoma with scattered erect setae.

Sculpture on head and mesosomal dorsum shagreened, surface dull to moderately shining. Pubescence dense on head, mesosomal dorsum, anterior petiolar scale, and gastric tergites 1–4. Pubescence often long and somewhat wooly in appearance. Mesopleura and metapleural bulla without pubescence and strongly shining.

Body and appendages concolorous testaceous to dark reddish brown.

Queen measurements: (n = 3) HL 0.96–1.05, HW 0.92–1.04, SL 0.88–0.94, FL 0.98–1.09, LHT 1.09–1.19, EL 0.34–0.40, MML 2.32–2.49, WL 8.57, CI 95–100, SI 90–96, OI 35–38, WI 34, FI 42–44.

Queen description: Large species (MML 2.32–2.49). Head slightly longer than broad to about as long as broad in full face view (CI 95–100), posterior margin straight to slightly concave. Eyes of moderate size (OI 35–38). Ocelli of moderate size. Antennal scapes of moderate length (SI 90–96), in full face view scapes in repose surpassing posterior margin by a length approximately that of first funicular segment.

Forewings long relative to mesosomal length (WI 34). Forewings with Rs+M subequal in length to M.f2. Legs moderately short relative to mesosomal length (FI 42–44).

Dorsum of mesosoma and metasoma with abundant fine erect to subdecumbent setae, mesoscutum with more than 20 fine suberect setae. Body and appendages concolorous medium brown.

Male measurements: (n = 5) HL 0.66–0.71, HW 0.63–0.78, SL 0.23–0.24, FL 0.91–1.03, LHT 0.90–1.02, EL 0.34–0.37, MML 1.44–1.57, WL 4.4–4.9, PH 0.32–0.38, CI 90–97, SI 32–35, OI 51–53, WI 29–31, FI 61–66.

Male diagnosis: Forewing with 2 submarginal cells; volsella with distal arm shorter than 1/3 length of proximal arm and shorter than cuspis; dorsal profile of volsella and proximal arm straight or only weakly concave; legs relatively short for Fuscum-group species (FI < 68).

Male description: Head slightly longer than broad in full face view (CI 90–97). Eyes relatively large (OI 51–53), occupying much of anterolateral surface of head and separated from posterolateral clypeal margin by a length less than width of antennal scape. Ocelli large and in full frontal view set above adjoining posterolateral margins. Antennal scape moderately long (SI 32–35), 85–100% length of 3rd antennal segment. Anterior clypeal margin broadly convex medially. Mandibles large and nearly worker-like, masticatory margin broad, much longer than inner margin, bearing 1–4 apical teeth followed by 8–14 denticles. Inner margin and exterior lateral margin diverging.

Mesosoma moderately developed, shorter in length than metasoma. Mesoscutum slightly enlarged, not projecting strongly forward or overhanging pronotum. Scutellum large, convex, nearly as tall as mesoscutum and projecting well above level of propodeum. Propodeum in lateral view not overhanging petiole, dorsal face rounding evenly into posterior face, posterior face straight to convex. Forewings long relative to mesosomal length (WI 29–31) and bearing two submarginal cells. Wing color clear to slightly smoky with darker brown veins and stigma. Legs long relative to mesosoma length (FI 61–66), although shorter than most Fuscum-group species.

Petiolar node bearing a blunt, broadly-rounded scale, node height taller than node length. Ventral profile of node strongly convex. Gaster elongate in dorsal view, 2.5–3 times as long as broad. Gonostylus produced as a slender filament. Volsella with ventrodistal process present as an elongate spine. Digitus elongate, distal arm short, shorter than 1/3 length of proximal arm and usually shorter than ventrodistal process. Cuspis absent. Proximal arm broad at base, greater than ½ height of adjoining volsella, and tapering distally.

Dorsal surfaces of body with scattered erect setae, mesoscutum with more than 10 erect setae. Venter of gaster with scattered setae. Pubescence dense on body and appendages, becoming sparse only on medial propodeal dorsum.

Head and body medium brown in color. Mandibles, antennae, and legs testaceous.

Distribution: Mountains of northwestern South America to Costa Rica.

Biology: Linepithema piliferum is a montane species, with records running from 780 to 2340 meters. Where explicit habitat information has been recorded, three collections are from sifted litter in wet forest, five from cloud forest edge or roadside, one from 2nd growth forest edge, and one from a *Psidium guayaba* orchard. Four nest records are from under stones and one from soil. Two port-of-entry intercepts into the United States were nesting in *Cattleya* orchids. This species has been recorded tending root aphids, aleyrodids, and pseudococcids.

Colonies can be populous and are probably both polydomous and polygynous. I conducted three nest excavations in Ecuador in December 2002. Alate males and queens were present in all nests. Each of these nests were under series of several stones along roadsides in cloud forest. Nests contained dozens of separate brood chambers connected by tunnels, and at least one nest contained a colony of root aphids feeding on grass roots growing through the nest. In two of the colonies I found several dealate queens, each in a separate brood chamber.

Similar species: Workers of *L. angulatum* usually have shorter antennal scapes (SI <104, Fig. 77), a more deeply impressed metanotal groove, and in South America lack standing setae on the cephalic dorsum (present in some Central American populations). Workers of *L. tsachila* have shorter antennal scapes (SI < 98), a more robust head in full-frontal view (CI usually > 94, Fig. 78) that typically reaches its widest point near the level of the compound eyes, a deeply concave posterior margin, and often a faint bluish sheen to the integument. Males of other Fuscum-group species either have a long distal arm of the digitus (*angulatum, fuscum,* and *keiteli*), or longer legs (FI > 70, *tsachila*, Fig. 88).

Discussion: This species can be difficult to diagnose owing to a large amount of intraspecific variation and the presence of two very similar sympatric species, *L. angulatum* and *L. tsachila*. All *L. piliferum* specimens have relatively long scapes and standing cephalic setae, but vary continuously in eye size, color, and degree of mesonotal impression. Specimens from Costa Rica, lower elevations in Ecuador, and parts of Colombia tend to have smaller eyes (< 60 ommatidia) and be lighter in color.

Crozier (1970) reports the karyotypes of "*Iridomyrmex pilifer*" and "*I.* sp. nr. *pilifer*", but Crozier's voucher specimens (MCZC, examined) reveal both ants to be *L. fuscum*.

Material examined: COLOMBIA. Cauca: Popayán, 1575m [WPMC]; Reserva Nat. El Guayaba, Popayán, 1600m [MCZC]. Chocó: Mpio. San José del Palmar, Vereda "El

Tabor", 1540m [UCDC]. Huila: Finca Merenberg, 12 km W Belén [MCZC]; Santa
Leticia [WPMC]. La Guajira: San Antonio (as "St. Antonio") [NHMB]. Nariño: 5 mi.
E. Santiago, 2160m [CASC]. Quindio: Filandia vda Cruces, Fca El Paraiso, 04°40'N
75°38'W, 1900m [UCDC]. Risaralda: Santa Rosa de Cabal [UCDC]. Tolima: El
Diamante (Chaparral) [WPMC]; Transecto Parque los Nevados, E slope, 2340m
[UCDC]. Valle: 11 mi. W Cali [CASC]; Bosque de Yotoco, 1575m [ALWC,
WPMC]; Medio Colima Site [WPMC]; NW Ibagué near Restrepo [WPMC]; Salidito,
W of Cali, TV Tower Road, 2100m [MCZC]; Sevilla [WPMC]. Colombia (s.loc.),
various Port-of-entry U.S. intercepts [LACM, USNM]. Colombia (s.loc.) [NHMW].
COSTA RICA. Alajuela: 14 km S Vol. Arenal, 10°20'N 84°43'W, 1000m [JTLC];
Peñas Blancas, 10°19'N 84°42'W, 780m [UCDC]; Rio Peñas Blancas, 10°19'N
84°43'W, 800m [JTLC]. ECUADOR. Napo: 11 km SE Cosanga, Road Baeza-
Archidona, 00°40'S 77°48'W, 1640m [ALWC, MCZC]; Cosanga, 00°35'S 77°51'W,
2000m [ALWC, MCZC, MZSP]. Pastaza: Mera [CASC]. Pichincha: 8 k West Baeza,
Road Quito to Baeza, 00°27'S 77°58'W, 2100m [ALWC, BMNH, CASC, IFML,
INBP, LACM, MCZC, MHNG, MZSP, NHMB, NHMW, PSWC, QCAZ, UCDC,
USNM, WPMC]; Las Pampas-Alluriquin Road, 00°23'S 78°59'W, 1300m [ALWC,
MZSP]; Nanegalito, 00°04'N 78°41'W, 1500m [ALWC]. VENEZUELA. Venezuela
(s.loc.), Port-of-entry U.S. intercept [USNM].

Linepithema pulex Wild, **sp. nov.**

(worker lateral Fig. 45; worker head Fig. 46; male head Fig. 69; distribution Fig. 105)

Iridomyrmex humilis Mayr st. *angulata* Em. var. *pulex* Santschi 1923: 68 (W).
Unavailable name (quadrinomen). [mat. ref. BRAZIL, Santa Catarina:
Blumenau, Reichensperger, 2 w, NHMB, examined]

Species group: unassigned

Holotype worker. PARAGUAY. Canindeyú: Reserva Natural del Bosque Mbaracayú,
Jejuimi, 24°08'S 055°32'W 12.xi.2002. Humid subtropical semideciduous forest, nest
1m above ground in litter between *Philodendron bipinnatifidum* vine and tree trunk,
and in tree bark. A.L.Wild acc. no. AW1678. [1 worker, INBP].

Paratypes. Same collection data as holotype, A. L. Wild acc. nos. AW1677–1678 [13
workers, 9 males, and 1 queen, ALWC, BMNH, CASC, IFML, LACM, MCZC,
MHNG, MZSP, NHMW, UCDC, USNM].

Holotype worker measurements: HL 0.57, HW 0.52, MFC 0.16, SL 0.48, FL 0.41,
LHT 0.38, ES 1.26, PW 0.32, CI 91, SI 92, CDI 31, OI 22.

Worker measurements: (n = 14) HL 0.48–0.65, HW 0.42–0.58, MFC 0.13–0.16, SL 0.42–0.54, FL 0.34–0.47, LHT 0.33–0.46, ES 0.82–1.76, PW 0.26–0.36, CI 86–95, SI 87–100, CDI 27–31, OI 16–27.

Worker diagnosis: Sculpture on head densely punctate-reticulate, surface dull and opaque; metanotal groove deeply impressed; propodeum raised; gastric tergites 3–4 with sparse pubescence; size small (HL < 0.65; FL < 0.47); antennal scapes short (SI 87–100); body color yellow to reddish brown.

Worker description: Head in full face view slightly longer than broad (CI 86–95), lateral margins convex, posterior margin concave. Compound eyes small to moderate (OI 16–27), comprised of 40–60 facets (usually < 50). Antennal scapes short (SI 87–100), shorter than head length. In full face view, scapes in repose surpass posterior margin of head by less then the length of the first funicular segment. Frontal carinae moderately to broadly spaced (CDI 27–31). Maxillary palps about ½ HL or less, ultimate segment longer than either of the preceding two segments, ultimate segment (segment six) longer than segment 2.

Mesosoma in lateral view with pronotum and mesonotum forming a single convexity, mesonotum relatively angular in most specimens, without central saddle or indentation. Mesonotal angle sometimes approaches 90° in larger workers. Metanotal groove deeply impressed, propodeum raised, dorsal face straight or with medial saddle-like impression, posterior margin concave.

Petiolar scale sharp and inclined anteriorly, in lateral view falling short of the propodeal spiracle.

Cephalic dorsum (excluding clypeus) either lacking erect setae or bearing a single pair of short setae near posterior margin. Pronotum with at least one pair of long erect anterior setae and often with a second or rarely a third shorter pair near the mesonotal suture. Mesonotum sometimes with a small pair of erect setae in the posterior quarter anterior of metanotal suture. Erect setae on gastric tergites 1–4 (= abdominal tergites 3–6) relatively abundant, tergite 1 bearing 3–8 erect setae (mean = 5.8). Venter of metasoma with scattered erect setae.

Sculpture on head and mesosomal dorsum densely and finely punctate, surface dull and opaque. Pubescence dense on head and mesosomal dorsum. Mesopleura and metapleura anterior of metapleural gland orifice without pubescence, surface glabrous. Gastric tergites 1–2 with moderate pubescence, becoming dilute on tergites 3–4, surface moderately shining.

Color testaceous to medium reddish brown, gaster often darker, light brown to dark brown.

Queen measurements: (n = 1) HL 0.72, HW 0.67, SL 0.57, FL 0.57, LHT 0.63, EL 0.23, MML 1.37, WL n/a, CI 94, SI 84, OI 32, WI n/a, FI 41.

Queen description: Small species (MML 1.37). Head longer than broad in full face view (CI 94), posterior margin slightly concave. Eyes moderately small (OI 32). Ocelli moderately small. Antennal scapes relatively short (SI 84), in full face view scapes in repose surpassing posterior margin by a length less than that of first funicular segment.

Wings unknown (not present on single examined queen). Legs short relative to mesosomal length (FI = 41).

Dorsum of mesosoma and metasoma with short, scattered standing setae. Mesoscutum with more than 10 standing setae. Body and appendages concolorous light reddish brown.

Male measurements: (n = 4) HL 0.45–0.47, HW 0.44–0.48, SL 0.11–0.12, FL 0.44–0.45, LHT 0.37–0.39, EL 0.17–0.19, MML 0.78–0.82, WL 2.21–2.26, PH 0.17–0.18, CI 97–103, SI 24–26, OI 37–41, WI 28–29, FI 55–58.

Male diagnosis: Forewing with 1 submarginal cell; eyes relatively small (EL < 0.20) propodeum with posterior face slightly concave, not strongly overhanging petiole; petiolar node bearing a tall dorsal scale; surface of head densely punctate and opaque.

Male description: Head about as broad as long in full face view (CI 97–103). Eyes relatively small (OI 38–41), occupying much of anterolateral surface of head anterior of midline and separated from posterolateral clypeal margin by a length less than width of antennal scape. Ocelli of moderate size and in full frontal view set above adjoining posterolateral margins. Antennal scape short (SI 24–26), about 2/3 length of 3rd antennal segment. Anterior clypeal margin convex medially. Mandibles small, masticatory margin broad, longer than inner margin, usually bearing a small apical tooth followed by a much longer subapical tooth and a series of 7–10 denticles. Inner margin and exterior lateral margin parallel to slightly diverging.

Mesosoma moderately developed and subequal in length to metasoma. Mesoscutum not greatly enlarged, projecting slightly forward over pronotum, dorsal margin in lateral view relatively flat. Scutellum equal in height to mesoscutum and well above level of propodeum. Propodeum in lateral view not strongly overhanging petiole, posterior face slightly concave. Forewings long relative to mesosomal length (WI 28–

29) and bearing one submarginal cell. Wings transparent, with pale brown wing veins and stigma. Legs of moderate length relative to mesosoma (FI 55–58).

Petiolar node bearing a blunt erect scale, node height taller than node length. Venter of node bearing a strong, posteriorly-projecting process. Gaster ovoid in dorsal view, about 2 times as long as broad. Gonostylus produced as a pointed, triangular pilose lobe. Volsella with cuspis present, digitus short and downturned distally.

Dorsal surfaces of body with scattered erect setae, mesoscutum with more than 4 erect setae. Venter of gaster with scattered setae. Pubescence dense on body and appendages, becoming sparse only on medial propodeal dorsum. Sculpture on head and mesoscutum finely and densely punctate, surface opaque.

Head dark brown. Mesosoma and metasoma pale brown to medium brown. Mandibles, antennae, and legs pale whitish-yellow.

Distribution: Southeastern Brazil, northeastern Argentina, and eastern Paraguay.

Biology: Linepithema pulex has been collected in a variety of habitats in the Atlantic coastal rainforest region, including mata atlântica, humid subtropical tall forest, forest edge, coffee plantation, and Restinga habitats. This species appears to be patchy in distribution and not commonly encountered (A. Wild pers. obs, R. Silva, pers. comm.) Nests of *L. pulex* were observed only at the type locality in Paraguay, where one superficial nest was in soil and leaf litter to a depth of 5cm, another nest fragment was found in the leaf litter, and the type series was found nesting 1m above ground in litter, at the junction between a small *Philodendron bipinnatifidum* Schott vine and tree trunk. This species has been recorded recruiting to tuna, sardine, and honey baits in Brazil and in Paraguay, and several records result from Winkler or Berlese litter sampling methods. Males were observed in the nest in November, and one nest was found to have a single dealate queen.

Similar species: Workers of *L. angulatum* are larger, the head less opaque, and the gaster with dense pubescence on all tergites. Workers of *L. cryptobioticum* have much shorter antennal scapes (SI < 85) and smaller eyes.

Discussion: Some Brazilian worker specimens have less extensive sculpture on the head, so that the sides of the head are relatively shining compared to Argentinean and Paraguayan specimens. A few specimens from Minas Gerais and São Paulo are darker in color than the type series. The shape of the mesonotum and the propodeum is also rather variable between localities, but this variation does not appear to show any broad geographic structure.

Santschi's (1923) name *pulex* is not available under ICZN rules (ICZN, 2000) as it was published as a quadrinomen. This study marks the first available use of *pulex* in *Linepithema*. Santschi's specimens from Santa Catarina (NHMB, examined) are clearly conspecific with the Paraguayan type series. The type series was chosen from a recent Paraguayan collection because Brazilian collection and specimen loan regulations render scientific study in that country difficult.

Etymology: Santschi's name *pulex* probably refers to the small size of this species.

Material examined: ARGENTINA. Misiones: Parque Nacional Iguazu, 25°42'S 54°26'W [AVSC]; Estación Experimental Loreto [IFML, MACN, NHMB]. BRAZIL. Espírito Santo: Santa Teresa [MCZC]; Pedro Canário Conc. de Barra [MZSP]. Goiás: Anápolis [MZSP]. Minas Gerais: Lagoa Dourada [MZSP]; Viçosa [MZSP]. Paraná: Iguaçu [MZSP]. Rio de Janeiro: Nova Iguaçu, Reserva Biológica do Tinguá, 22°34'S; 43°24'W 400m [ALWC]. Santa Catarina: Blumenau [MZSP, NHMB]; Gaspar [MZSP]. São Paulo: Cunha, P.E. Serra do Mar, Nucléo Cunha-Indaia, 23°15'S 45°00'W [MZSP]; Iguape, E.E. Jureia-Itatins, Nucléo Rio Verde, 24°33'S 47°14'W [MZSP]; Lençois [MZSP]; Praia Grande, P.E. Serra do Mar, Nucléo Pilões-Cubatão, 23°59'S 46°32'W [MZSP]; V. Estr. Santos, Meio da Serra [MZSP]. PARAGUAY. Canindeyú: Res. Nat. del Bosque Mbaracayú, Jejuimi, 24°08'S 55°26'W [ALWC, BMNH, CASC, IFML, LACM, MCZC, MHNG, MZSP, NHMW, UCDC, USNM].

Linepithema tsachila Wild, **sp. nov.**

(worker mesosoma Fig. 19; worker head Fig. 20; male wing Fig. 47; male body Fig. 51; male head Fig. 63; male volsella Fig. 64; distribution 103)

Species group: Fuscum

Holotype worker. ECUADOR. Pichincha: ENDESA forest, 00°06'N 79°02'W, 700m, 5.xii.2003, Next to station buildings in mixed tree plantation, Trail up tree trunk in old termite tubes. A. L. Wild acc. no. AW2212 [1 worker, QCAZ].

Paratypes. Same data as holotype [26 workers and 8 males, ALWC, BMNH, LACM, MCZC, MHNG, MZSP, UCDC, USNM].

Holotype worker measurements: HL 0.66, HW 0.63, MFC 0.17, SL 0.57, FL 0.52, LHT 0.54, ES 1.50, PW 0.40, CI 94, SI 91, CDI 27, OI 23.

Worker measurements: (n = 13) HL 0.53–0.72, HW 0.49–0.72, MFC 0.14–0.19, SL 0.53–0.65, FL 0.45–0.62, LHT 0.46–0.62, ES 1.08–2.14, PW 0.31–0.45, CI 86–99, SI 87–97, CDI 25–29, OI 19–29.

Worker diagnosis: Posterior margin of head in full face view concave; head broad, reaching widest point near level of compound eyes; cephalic dorsum with a pair of standing setae near posterior margin; mesonotum without strong medial impression; mesopleura and metapleura lacking pubescence and moderately shining; integument often with a faint blue sheen.

Worker description: Head in full face view relatively broad (CI 86–99, usually > 94), lateral margins convex, running subparallel or converging posterior of compound eyes. Posterior margin slightly to deeply concave. Head in most workers reaches widest point near level of compound eyes. Compound eyes of moderate size (OI 19–29), comprised of 45–60 ommatidia. Antennal scapes relatively short (SI 87–97), shorter than head length. In full face view, scapes in repose exceed posterior margin of head by a length less than length of first funicular segment. Frontal carinae moderately spaced (CDI 25–29). Maxillary palps of moderate length, approximately ½ HL or less, ultimate segment (segment 6) shorter than or subequal in length to segment 2.

Pronotum and anterior mesonotum in lateral view forming a single convexity. Mesonotal dorsum relatively straight, sometimes with a slight mesal impression. Metanotal groove not impressed or only slightly impressed. Propodeum in lateral view relatively high, dorsal propodeal face straight to convex, posterior propodeal face convex.

Petiolar scale relatively sharp and inclined anteriorly, in dorsal view relatively narrow, in lateral view falling short of propodeal spiracle.

Cephalic dorsum (excluding clypeus) with a pair of suberect to subdecumbent setae near vertex and sometimes with 1–3 additional standing hairs closer to antennal insertions. Pronotum with 2–8 erect to suberect setae (mean = 4.9), some setae longer than maximum diameter of compound eye. Mesonotum without erect setae. Gastric tergite 1 (= abdominal tergite 3) bearing 1–4 erect setae (mean = 1.9) mesally, exclusive of a row of 6–15 subdecumbent setae along posterior margin of tergite, tergite 2 bearing 2–6 erect setae (mean = 4.1) exclusive of posterior row, tergite 3 bearing 5–7 erect setae (mean = 6.0) exclusive of posterior row. Venter of metasoma with scattered erect setae.

Sculpture on head and mesosomal dorsum shagreened and relatively opaque. Pubescence dense on head, mesosomal dorsum, and gastric tergites 1–4. Mesopleura and metapleura without pubescence and shining.

Color testaceous to medium brown, head often somewhat darker than mesosoma. Surface of body sometimes with a faint opaque bluish sheen.

Queen measurements: (n = 4) HL 0.92–1.02, HW 0.92–0.97, SL 0.78–0.85, FL 0.90–1.00, LHT 1.00–1.14, EL 0.36–0.47, MML 1.97–2.28, WL 6.85–7.04, CI 93–100, SI 85–89, OI 39–46, WI 31–32, FI 42–46.

Queen description: Moderately large species (MML 1.97–2.28). Head slightly longer than broad to about as long as broad in full face view (CI 93–100), posterior margin concave. Eyes unusually large (OI 39–46). Ocelli large. Antennal scapes relatively short (SI 85–89), in full face view scapes in repose surpassing posterior margin by a length less than that of first funicular segment.

Forewings moderately long relative to mesosomal length (WI 31–32). Forewings with Rs+M somewhat greater in length than M.f2. Legs moderately short relative to mesosomal length (FI 42–46).

Dorsum of mesosoma and metasoma with scattered erect to subdecumbent setae, mesoscutum with more than 10 suberect to subdecumbent setae. Body and appendages concolorous medium to brown.

Male measurements: (n = 4) HL 0.71–0.74, HW 0.69–0.71, SL 0.21–0.24, FL 1.10–1.15, LHT 1.09–1.13, EL 0.39–0.43, MML 1.53–1.62, WL 4.35–4.51, PH 0.29–0.33, CI 93–97, SI 30–32, OI 55–58, WI 27–28, FI 70–71.

Male diagnosis: Forewing with 2 submarginal cells; volsella with distal arm shorter than 1/3 length of proximal arm; dorsal profile of volsella and proximal arm strongly concave; eyes large (EL > 0.38, OI > 54).

Male description: Head slightly longer than broad in full face view (CI 93–97). Eyes large (OI 55–58), occupying much of anterolateral surface of head and separated from posterolateral clypeal margin by a length less than width of antennal scape. Ocelli very large, median ocellus as wide as separation between antennal insertions. Antennal scape moderately long (SI 30–32), 75–90% length of 3rd antennal segment. Anterior clypeal margin broadly convex medially. Mandibles large and worker-like, masticatory margin broad, much longer than inner margin, bearing 1–2 apical teeth followed by alternating series of teeth and denticles, similar to worker dentition. Inner margin and exterior lateral margin strongly diverging.

Mesosoma moderately developed, shorter in length than metasoma. Mesoscutum slightly enlarged, not projecting strongly forward or overhanging pronotum. Scutellum large, convex, nearly as tall as mesoscutum and projecting well above level of propodeum. Propodeum in lateral view not overhanging petiole, dorsal face rounding evenly into posterior face, posterior face straight to convex. Forewings long relative to mesosomal length (WI 27–28) and bearing two submarginal cells. Wing color clear to slightly smoky with darker brown veins and stigma. Legs long relative to mesosoma length (FI 70–71).

Petiolar node bearing a blunt, broadly-rounded scale, node height slightly shorter than node length. Ventral profile of node straight to slightly convex. Gaster elongate in dorsal view, 3–3.5 times as long as broad. Gonostylus produced as a slender filament. Volsella with ventrodistal process present as a spine or sharp tooth. Digitus elongate, distal arm short and hook-like, shorter than 1/3 length of proximal arm. Proximal arm broad at base, greater than ½ height of adjoining volsella, and tapering distally. Dorsal profile of volsella and proximal arm strongly concave, digitus arching upwards. Cuspis absent.

Dorsal surfaces of body with scattered erect setae, mesoscutum with more than 4 erect setae. Venter of gaster with scattered setae. Pubescence dense on body and appendages, becoming sparse only on medial propodeal dorsum.

Head and body medium brown in color. Mandibles, antennae, and legs lighter.

Distribution: Colombia, Ecuador.

Biology: Linepithema tsachila is locally abundant along the western slopes of the Andes in Ecuador and Colombia with collection records running from near sea level to 2,000 meters. This species occurs in a broad range of habitats and can reach high numbers in some human disturbed areas. Eight collection records are from rainforest, seven from tree plantations or orchards, five from forest edge habitats, four from pastures, and three from second growth forest. Six nesting records are from under stones, along with one record each from rotting wood, soil, and leaf litter. I observed several colonies in Ecuador in 2002. Surface nest entrances are inconspicuous and nests are often diffuse, spreading out as networks of tunnels under several rocks and in the soil and litter. One colony was observed making use of old termite carton tubes. *L. tsachila* has been recorded tending honeydew-producing insects, once with coccids and once with pseudococcids on a *Cecropia* tree, and in Ecuador I observed these ants scavenging pieces of dead insects, recruiting to an earthworm, and recruiting to the body of a recently-killed rodent. Males have been taken at light traps in August and alates collected in nests in May and December.

Similar species: Workers of the sister species *L. piliferum*, whose range broadly overlaps with that of *L. tsachila*, have longer antennal scapes (SI > 99, Fig. 77), a narrower head (Fig. 78), and lack the faint opaque bluish sheen to the integument. Workers of *L. angulatum* (in South America) and *L. fuscum* lack standing setae on the vertex and usually also have fewer pronotal setae. Additionally, *L. fuscum* workers are normally somewhat smaller (HW < 0.52). Males of other Fuscum-group species have smaller ocelli and a much less concave dorsal margin of the volsella.

Discussion: Linepithema tsachila shows habitat and nesting associations much like those of the morphologically similar sister species, *L. piliferum*. The two species are locally sympatric in some localities in Ecuador with no signs of hybridization, an observation that supports the present treatment of *L. tsachila* as a valid species.

Most populations of this species are uniform in appearance. At Unión Toachí, at 1000 meters on the western Andean slopes, some collections are smaller and darker in color with relatively longer scapes. These might represent workers from young colonies, as molecular evidence indicates little or no DNA sequence divergence in several loci from large, light colored forms collected at the same location (Wild, unpublished data).

Etymology: This name refers to the Tsa'chila people of western Ecuador, and is given as a noun in apposition.

Material examined: COLOMBIA. Chocó: 10 km SW S. Jose del Palmar, Rio Torito, 760m [MCZC]. Valle: 3.2 km E. Rio Aguaclara on old Cali Rd [MCZC]. ECUADOR. Los Rios: Corina, nr. Patricia Pilar [UCDC]; Rio Palenque Sta., 47 km S Santo Domingo, 215m [MCZC]; Rio Palenque, 2 km SSE Patricia Pilar, 160m [MCZC, UCDC]. Orellana: Yasuni Scientific Research Station, 250m [ALWC, UCDC]. Pichincha: ENDESA forest, 00°06'N 79°02'W, 700m [ALWC, BMNH, LACM, MCZC, MHNG, MZSP, QCAZ, UCDC, USNM]; ENDESA forest , 00°08'N 79°03'W, 600m [ALWC, CASC, MCZC, NHMB, UCDC]; ENDESA forest concession [MCZC]; Las Pampas-Alluriquin Road, 00°24'S 78°59'W, 1500m [ALWC, IFML, LACM]; Maquipucuna, 5 km ESE Nanegal, 00°07'N 78°38'W, 1250m [MCZC, PSWC]; Road to Mindo, 00°02'S 78°46'W, 1525m [ALWC, UCDC]; Nanegalito, 00°04'N 78°41'W, 1500m [ALWC, BMNH]; Unión Toachi, 00°19'S 78°57'W, 1000m [ALWC, LACM, MZSP, QCAZ, UCDC, USNM].

LITERATURE CITED

Bernard, F. 1967. Faune de l'Europe et du Bassin Méditerranéen. 3. Les fourmis (Hymenoptera Formicidae) d'Europe occidentale et septentrionale. Paris: Masson, 411 pp.

Bolton, B. 1994. Identification Guide to the Ant Genera of the World. Harvard University Press, Cambridge, MA, 222 pp.

Bolton, B. 1995. A new general catalogue of the ants of the world. Harvard University Press, Cambridge, MA, 504 pp.

Borgmeier, T. 1928. Algumas formigas do Museu Paulista. Bol. Biol. Lab. Parasitol. Fac. Med. São Paulo 12: 55-70.

Brandão, C. R. F., Baroni Urbani, C., and Wagensberg, J. 1999. New *Technomyrmex* in Dominican amber (Hymenoptera: Formicidae), with a reappraisal of Dolichoderinae phylogeny. Entomol. Scand. 29: 411-428.

Brèthes, J. 1914. Note sur quelques Dolichodérines argentines. An. Mus. Nac. Hist. Nat. B. Aires 26: 93-96.

Brown, W. L. 1958. A review of the ants of New Zealand. Acta Hymenopterol. 1: 1-50.

Brown, W. L., and Nutting, W. L. 1950. Wing venation and the Phylogeny of the Formicidae. Trans. Am. Entomol. Soc. 75: 113-132.

Buczkowski, G., Vargo, E. L., Silverman, J. 2004. The diminutive supercolony: the Argentine ants of the southeastern United States. Molecular Ecology 13: 2235-2242.

Cavill, G. W. K. and Houghton, E. 1973. Hydrocarbon constituents of the Argentine ant, *Iridomyrmex humilis*. Australian Journal of Chemistry 26: 1131-1135.

Cavill, G. W. K. and Houghton, E. 1974. Volatile constituents of the Argentine ant, *Iridomyrmex humilis*. Journal of Insect Physiology 20: 2049-2059.

Cavill, G. W. K., Davies, N.W. and McDonald, F.J. 1980. Characterization of aggregation factors and associated compounds from the Argentine ant, *Iridomyrmex humilis*. Journal of Chemical Ecology 6: 371-384.

Chapin, F. S., Zavaleta, E. S., Eviner, V. T., Naylor, R. L., Vitousek, P. M., Reynolds, H. L., Hooper, D. U., Lavorel, S., Sala, O. E., Hobbie, S. E., Mack, M. C., and Diaz, S. 2000. Consequences of changing biodiversity. Nature 405: 234-242.

Chiotis, M., Jermiin, L. S. and Crozier, R. H.. 2000. A molecular framework for the phylogeny of the ant subfamily Dolichoderinae. Mol. Phylogenet. Evol. 17(1): 108-116.

Chopard, L. 1921. La fourmi d'Argentine *Iridomyrmex humilis* var. *arrogans* Santschi dans le midi de la France. Ann. Epiphyt. 7: 237-265.

Christian, C. E. 2001. Consequences of a biological invasion reveal the importance of mutualism for plant communities. Nature 413: 635-639.

Coyne, J. A. and Orr, H. A. 2004. Speciation. Sinauer, Sunderland, MA, 557 pp.
Dalla Torre, K. W. von. 1893. Catalogus Hymenopterorum hucusque
    descriptorum systematicus et synonymicus. Vol. 7. Formicidae (Heterogyna).
    Leipzig: W. Engelmann, 289 pp.
Creighton, W. S. 1950. The ants of North America. Bull. Mus. Comp. Zool. 104: 1-
    585.
Crozier, R. H. 1970. Karyotypes of twenty-one ant species (Hymenoptera:
    Formicidae), with reviews of the known ant karyotypes. Can. J. Genet. Cytol.
    12: 109-128.
Emery, C. 1888. Über den sogenannten Kaumagen einiger Ameisen. Z. Wiss. Zool.
    46: 378-412.
Emery, C. 1890. Studii sulle formiche della fauna neotropica. Bull. Soc. Entomol. Ital.
    22: 38-80.
Emery, C. 1894a. Studi sulle formiche della fauna neotropica. VI-XVI. Bull. Soc.
    Entomol. Ital. 26: 137-241.
Emery, C. 1894b. [Untitled.] Pp. 373-401 in: Ihering, H. von. Die Ameisen von Rio
    Grande do Sul. Berl. Entomol. Z. 39: 321-447.
Emery, C. 1894c. Estudios sobre las hormigas de Costa Rica. An. Mus. Nac. Costa
    Rica 1888-1889: 45-64.
Emery, C. 1912. Hymenoptera. Fam. Formicidae. Subfam. Dolichoderinae. Genera
    Insectorum. 137 : 1-50.
Forel, A. 1885. Études myrmécologiques en 1884 avec une description des organs
    sensoriels des antennes. Bull. Soc. Vaudoise Sci. Nat. 20: 316-380.
Forel, A. 1901. I. Fourmis mexicaines récoltées par M. le professeur W.-M. Wheeler.
    II. A propos de la classification des fourmis. Ann. Soc. Entomol. Belg. 45:
    123-141.
Forel, A. 1907. Formiciden aus dem Naturhistorischen Museum in Hamburg. II. Teil.
    Neueingänge seit 1900. Mitt. Naturhist. Mus. Hambg. 24: 1-20.
Forel, A. 1908. Ameisen aus Sao Paulo (Brasilien), Paraguay etc. gesammelt von
    Prof. Herm. v. Ihering, Dr. Lutz, Dr. Fiebrig, etc. Verh. K-K. Zool.-Bot. Ges.
    Wien 58: 340-418.
Forel, A. 1911. Ameisen des Herrn Prof. v. Ihering aus Brasilien (Sao Paulo usw.)
    nebst einigen anderen aus Südamerika und Afrika (Hym.). Dtsch. Entomol. Z.
    1911: 285-312.
Forel, A. 1912. Formicides néotropiques. Part V. 4me sous-famille Dolichoderinae.
    Mém. Soc. Entomol. Belg. 20: 33-58.
Forel, A. 1913. Fourmis d'Argentine, du Brésil, du Guatémala & de Cuba reçues de
    M. M. Bruch, Prof. v. Ihering, Mlle Baez, M. Peper et M. Rovereto. Bull. Soc.
    Vaudoise Sci. Nat. 49: 203-250.
Forel, A. 1914. Formicides d'Afrique et d'Amérique nouveaux ou peu connus. Bull.
    Soc. Vaudoise Sci. Nat. 50: 211-288.
Funk D. J. and Omland, K. E. 2003. The frequency, causes and consequences of

species level paraphyly and polyphyly: insights from animal mitochondrial DNA. Ann. Rev. Ecol. Syst. 34: 397-423.

Gallardo, A. 1916. Las hormigas de la República Argentina. Subfamilia Dolicoderinas. An. Mus. Nac. Hist. Nat. B. Aires 28: 1-130.

Giraud, T., Pedersen, J. S., and Keller, L. 2002. Evolution of supercolonies: The Argentine ants of southern Europe. Proc. Natl. Acad. Sci. USA 99: 6075 6079.

Goetsch, W. 1957. The ants. [English translation of "Die Staaten der Ameisen", 2nd edition.] Ann Arbor: University of Michigan Press, 173 pp.

Gotwald, W. H., Jr. 1995. Army ants: the biology of social predation. Ithaca, New York: Cornell University Press, xviii + 302 pp.

Harris, R. J. 2002. Potential impact of the Argentine ant (*Linepithema humile*) in New Zealand and options for its control. Science for Conservation 196, 36 pp.

Heller, N. E. 2004. Colony structure in introduced and native populations of the invasive Argentine ant, *Linepithema humile*. Insect. Soc. 51: 378-386.

Holway, D. A., and Suarez, A. V. 2004. Colony structure variation and interspecific competitive ability in the invasive Argentine ant. Oecologia 138: 216-222.

Human, K. G., and Gordon, D. M. 1997. Effects of Argentine ants on invertebrate biodiversity in northern California. Conservation Biology 11: 1242-1248.

Kempf, W. W. 1969. Miscellaneous studies on Neotropical ants. V. (Hymenoptera, Formicidae). Stud. Entomol. 12: 273-296.

Ketterl, J., Verhaagh, M., Bihn, J. H., Brandão, C. R. F., and Engels, W. 2003. Spectrum of ants associated with *Araucaria angustifolia* trees and their relations to Hemipteran trophobionts. Stud. Neotrop. Fauna and Environ. 38: 199-206.

Krieger, M. J. B., and Keller, L. 2000. Mating frequency and genetic structure of the Argentine ant *Linepithema humile*. Mol. Ecol. 9: 119-126.

Kusnezov, N. 1969. Nuevas especies de hormigas. Acta Zool. Lilloana 24: 33-38. Luederwaldt, H. 1926. Observações biologicas sobre formigas brasileiras especialmente do estado de São Paulo. Rev. Mus. Paul. 14: 185-303.

Mattos, M. R., and Orr, M. R. 2002. Two new *Pseudacteon* species (Diptera: Phoridae), parasitoids of ants of the genus *Linepithema* (Hymenoptera: Formicidae) in Brazil. Stud. Dipt. 91: 283-288.

Mayr, E. 1942. Systematics and the origin of species from the viewpoint of a zoologist. New York, Columbia University Press, 334 pp.

Mayr, G. 1866. Myrmecologische Beiträge. Sitzungsber. Kais. Akad. Wiss. Wien Math.-Naturwiss. Cl. Abt. I 53: 484-517.

Mayr, G. 1868. Formicidae novae Americanae collectae a Prof. P. de Strobel. Annu. Soc. Nat. Mat. Modena 3: 161-178.

Mayr, G. 1870a. Formicidae novogranadenses. Sitzungsber. Kais. Akad. Wiss. Wien Math.-Naturwiss. Cl. Abt. I 61: 370-417.

Mayr, G. 1870b. 1870 Neue Formiciden. Verh. K-K. Zool.-Bot. Ges. Wien 20: 939-

996.

Menozzi, C. 1935. Fauna Chilensis. II. (Nach Sammlungen von W. Goetsch). Le formiche del Cile. Zool. Jahrb. Abt. Syst. Ökol. Geogr. Tiere 67: 319-336.

Newell, W., and Barber, T. C. 1913. The Argentine ant. Bulletin (United States. Bureau of Entomology); no. 122. Washington, D.C. : U.S. Dept. of Agriculture, Bureau of Entomology, 98 p.

Orr, M. R., De Camargo, R. X., and Benson, W. W. 2003. Interactions between ant species increase arrival rates of an ant parasitoid Anim. Behav. 65: 1187-1193.

Orr, M. R., and Seike, S. H. 1998. Parasitoids deter foraging by Argentine ants (*Linepithema humile*) in their native habitat in Brazil Oecologia 117 : 420-425.

Orr, M. R., Seike, S. H., Benson, W. W., and Dahlsten, D. L. 2001. Host specificity of *Psuedacteon* (Diptera: Phoridae) Parasitoids that attack *Linepithema* (Hymenoptera: Formicidae) in South America Environ. Entomol. 30: 742-747.

Palacio, E. E., and Fernández, F. (ed.). 2003. "Clave para las subfamilies y géneros", pg. 233-260, in Fernández, F. (ed.). 2003. Introducción a las Hormigas de la región Neotropical. Instituto de Investigación de Recursos Biológicos Alexander Von Humboldt, Bogotá, Colombia, 424 pp.

Pressick, M. L., and Herbst, E. 1973. Distribution of ants on St. John, Virgin Islands. Caribb. J. Sci. 13: 187-197.

Reuter, M., Balloux, F., Lehmann, L., and Keller, L. 2001. Kin structure and queen execution in the Argentine ant *Linepithema humile*. Journal of Evolutionary Biology 14, 954-958.

Roura-Pascual, N., Suarez, A. V., Gomez, C., Pons, P., Touyama, Y., Wild, A. L. and Peterson, A. T. 2004. Geographical potential of Argentine ants (*Linepithema humile* Mayr) in the face of global climate change. Proc. Roy. Soc. London Ser. B 271: 2527-2535.

Santschi, F. 1916. Formicides sudaméricains nouveaux ou peu connus. Physis (B. Aires) 2: 365-399.

Santschi, F. 1919. Nouveaux formicides de la République Argentine. An. Soc. Cient. Argent. 87: 37-57.

Santschi, F. 1923. *Pheidole* et quelques autres fourmis néotropiques. Ann. Soc. Entomol. Belg. 63: 45-69.

Santschi, F. 1929. Nouvelles fourmis de la République Argentine et du Brésil. An. Soc. Cient. Argent. 107: 273-316.

Shattuck, S. O. 1992a. Review of the dolichoderine ant genus *Iridomyrmex* Mayr with descriptions of three new genera (Hymenoptera: Formicidae). J. Aust. Entomol. Soc. 31: 13-18.

Shattuck, S. O. 1992b. Higher classification of the ant subfamilies Aneuretinae,

Dolichoderinae and Formicinae (Hymenoptera: Formicidae). Syst. Entomol. 17: 199-206.

Shattuck, S. O. 1992c. Generic revision of the ant subfamily Dolichoderinae (Hymenoptera: Formicidae). Sociobiology 21: 1-181.

Shattuck, S. O. 1994. Taxonomic catalog of the ant subfamilies Aneuretinae and Dolichoderinae (Hymenoptera: Formicidae). Univ. Calif. Publ. Entomol. 112: i-xix, 1-241.

Shattuck, S. O. 1995. Generic-level relationships within the ant subfamily Dolichoderinae (Hymneoptera: Formicidae). Syst. Entomol. 20: 217-228.

Smith, M. R. 1929. Two introduced ants not previously known to occur in the United States. J. Econ. Entomol. 22: 241-243.

Smith, M. R. 1942. The relationship of ants and other organisms to certain scale insects on coffee in Puerto Rico. J. Agri. Univ. Puerto Rico 26: 21-27.

Snelling, R. R., and Hunt, J. H. 1975. The ants of Chile (Hymenoptera: Formicidae). Rev. Chil. Entomol. 9: 63-129.

Suarez, A. V., and Case, T. J. 2003. The ecological consequences of a fragmentation mediated invasion: The Argentine Ant, *Linepithema humile*, in southern California. Pages 161-180 in G.A. Bradshaw and P. Marquet (eds.) How landscapes change: Human disturbance and ecosystem disruptions in the Americas. Ecological Studies, vol. 162. Springer Verlag, Berlin.

Suarez, A. V., Holway, D. A., and Case, T. J. 2001. Patterns of spread in biological invasions dominated by long-distance jump dispersal: Insights from Argentine ants. Proc. Natl. Acad. Sci. USA 98: 1095-1100.

Touyama, Y., Ogata, K., and Sugiyama, T. 2003. The Argentine ant, *Linepithema humile*, in Japan: Assessment of impact on species diversity of ant communities in urban environments. Entomological Science 6, 57-62.

Tsutsui, N. D., and Case, T. J. 2001. Population genetics and colony structure in the Argentine ant (*Linepithema humile*) in its native and introduced ranges. Evolution 55: 976-985.

Ward, P. S., and Downie, D. A. 2005. The ant subfamily Pseudomyrmecinae (Hymenoptera: Formicidae): phylogeny and evolution of big-eyed arboreal ants. Syst. Entomol. 30: 310-335.

Wheeler, G. C., and Wheeler, J. 1951. The ant larvae of the subfamily Dolichoderinae. Proc. Entomol. Soc. Wash. 53: 169-210.

Wheeler, G. C., and Wheeler, J. 1974. Ant larvae of the subfamily Dolichoderinae: second supplement (Hymenoptera: Formicidae). Pan-Pac. Entomol. 49: 396-401.

Wheeler, Q. D., and Meier, R. (eds.). 2000. Species Concepts and Phylogenetic Theory: A Debate. New York, Columbia Univ. Press, 256 pp.

Wheeler, W. M. 1908. The ants of Porto Rico and the Virgin Islands. Bull. Am. Mus. Nat. Hist. 24: 117-158.

Wheeler, W. M. 1913a. [Untitled. Description of *Iridomyrmex humilis* Mayr.] Pp. 27

29 in: Newell, W., Barber, T. C. The Argentine ant. U. S. Dep. Agric. Bur. Entomol. Bull. 122: 1-98.

Wheeler, W. M. 1913b. Ants collected in the West Indies. Bull. Am. Mus. Nat. Hist. 32: 239-244.

Wheeler, W. M. 1929. Two Neotropical ants established in the United States. Psyche (Camb.) 36: 89-90.

Wheeler, W. M. 1942. Studies of Neotropical ant-plants and their ants. Bull. Mus. Comp. Zool. 90: 1-262.

Wheeler, W. M., and Mann, W. M. 1914. The ants of Haiti. Bull. Am. Mus. Nat. Hist. 33: 1-61.

Wild, A. L. 2004. Taxonomy and distribution of the Argentine ant, *Linepithema humile* (Hymenoptera: Formicidae). Ann. Entomol. Soc. Am. 97: 1204-1215.

Wild, A. L. and Cuezzo, F. 2006. Rediscovery of a fossil dolichoderine ant lineage (Hymenoptera: Formicidae: Dolichoderinae) and a description of a new genus from South America. Zootaxa 1142: 57-68.

Wilson, E. O. 1985. Ants of the Dominican amber (Hymenoptera: Formicidae). 3. The subfamily Dolichoderinae. Psyche 92: 17-37.

FIGURES

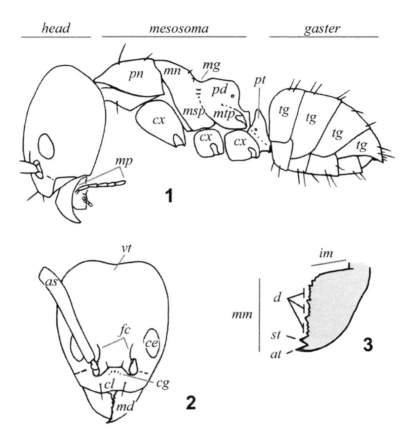

Figures 1-3. Schematic drawings of a *Linepithema* worker illustrating relevant morphological features. 1. Body, lateral view. 2. Head, full face view. 3. Mandible, dorsal view. Abbreviations are as follows: *as*, antennal scape; *at*, apical tooth of mandible; *ce*, compound eye; *cg*, pre-sutural clypeal groove; *cl*, clypeus; *cx*, coxa; *d*, denticles; *fc*, frontal carinae; *im*, inner margin of mandible; *md*, mandible; *mg*, metanotal groove; *mm*, masticatory margin of mandible; *mn*, mesonotum; *mp*, maxillary palps; *msp*, mesopleuron; *mtp*, metapleuron; *pd*, propodeum; *pn*, pronotum; *pt*, petiole; *st*, subapical tooth of mandible; *tg*, gastric tergite; *vt*, vertex.

Figures 4-6. Lateral views of *Linepithema* worker mesosoma
showing variation in mesopleural and metapleural pubescence.

4. *Linepithema dispertitum*, specimen from 3 km SSE San Andrés
    Semetabaj, Sololá, Guatemala.
5. *Linepithema micans*, specimen from Costanera Sur, Buenos Aires,
    Argentina.
6. *Linepithema neotropicum*, specimen from Archidona, Napo,
    Ecuador.

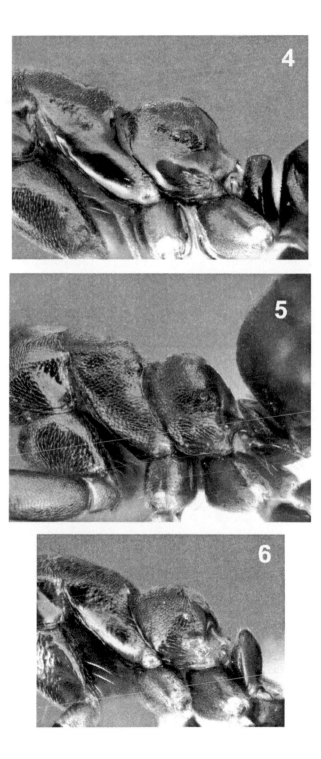

Figures 7-20. Line drawings of Fuscum-group worker specimens. Head and mesosoma figures of conspecifics are drawn from the same specimens.

7. *Linepithema angulatum*, mesosoma and petiole, lateral view, specimen from 2 km E Rio Negro, Tungurahua, Ecuador.
8. *Linepithema angulatum*, head, full face view.
9. *Linepithema cryptobioticum* holotype, mesosoma and petiole, lateral view.
10. *Linepithema cryptobioticum* holotype, head, full face view.
11. *Linepithema flavescens*, mesosoma and petiole, lateral view, specimen from La Hotte NE foothills, Haiti.
12. *Linepithema flavescens*, head, full face view.
13. *Linepithema fuscum*, mesosoma, lateral view, specimen from 15 km NE Pto. Maldonado, Madre de Dios, Peru.
14. *Linepithema fuscum*, head, full face view.
15. *Linepithema keiteli*, mesosoma, lateral view, specimen from Las Abejas, Pedernales, Dominican Republic.
16. *Linepithema keiteli*, head, full face view.
17. *Linepithema piliferum*, mesosoma, lateral view, specimen from Santa Rosa de Cabal, Risaralda, Colombia.
18. *Linepithema piliferum*, head, full face view.
19. *Linepithema tsachila* holotype, mesosoma, lateral view.
20. *Linepithema tsachila* holotype, head, full face view.

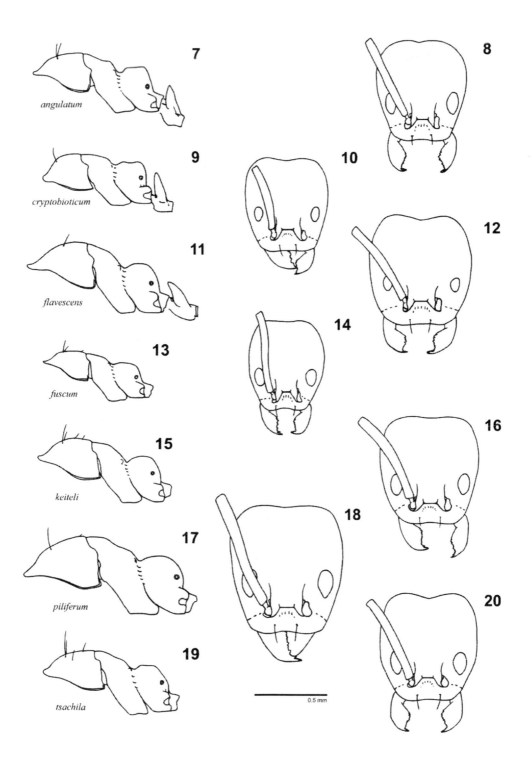

Figures 21-30. Line drawings of Iniquum-group and Neotropicum-group worker specimens. Head and mesosoma figures of conspecifics are drawn from the same specimens.

21. *Linepithema dispertitum*, mesosoma, lateral view, specimen from 3 km SSE San Andrés Semetabaj, Sololá, Guatemala.
22. *Linepithema dispertitum*, head, full face view.
23. *Linepithema iniquum*, mesosoma and petiole, lateral view, specimen from Baños, Tungurahua, Ecuador.
24. *Linepithema iniquum*, head, full face view.
25. *Linepithema leucomelas* lectotype, mesosoma and petiole, lateral view.
26. *Linepithema leucomelas* lectotype, head, full face view.
27. *Linepithema cerradense* holotype, mesosoma, lateral view.
28. *Linepithema cerradense* holotype, head, full face view.
29. *Linepithema neotropicum* holotype, mesosoma, lateral view.
30. *Linepithema neotropicum* holotype, head, full face view.

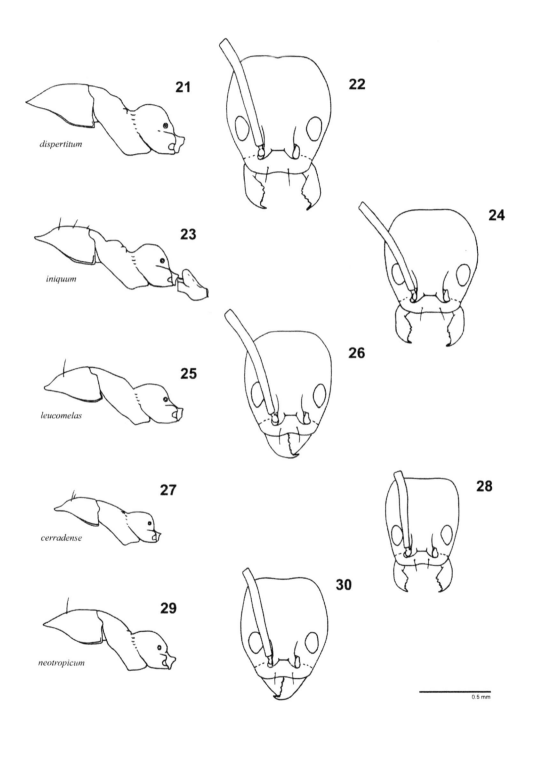

21

dispertitum

22

23

iniquum

24

25

leucomelas

26

27

cerradense

28

29

neotropicum

30

0.5 mm

Figures 31-42. Line drawings of Humile-group worker specimens. Head and mesosoma figures of conspecifics are drawn from the same specimens.

31. *Linepithema humile*, body, lateral view, specimen from Ñeembucú, Paraguay.
32. *Linepithema humile*, head, full face view.
33. *Linepithema anathema* holotype, mesosoma and petiole, lateral view.
34. *Linepithema anathema* holotype, head, full face view.
35. *Linepithema gallardoi*, mesosoma and petiole, lateral view, specimen from La Falda, Córdoba, Argentina.
36. *Linepithema gallardoi*, head, full face view.
37. *Linepithema micans* lectotype, mesosoma, lateral view.
38. *Linepithema micans* lectotype, head, full face view.
39. *Linepithema micans* (*platense* Forel lectotype), mesosoma, lateral view.
40. *Linepithema micans* (*platense* Forel lectotype), head, full face view.
41. *Linepithema oblongum*, mesosoma, lateral view, specimen from 7 km N Tafi de Valle, Tucumán, Argentina.
42. *Linepithema oblongum*, head, full face view.

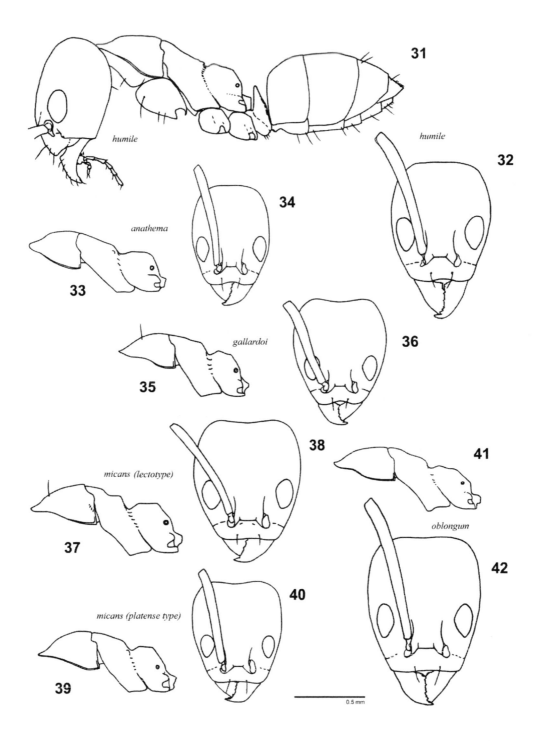

Figures 43-46. Line drawings of *Linepithema* worker specimens.
Head and mesosoma figures of conspecifics are drawn from the same
specimens.

43. *Linepithema aztecoides* holotype, body, lateral view.
44. *Linepithema aztecoides*, head, full face view.
45. *Linepithema pulex*, body, lateral view, specimen from Reserva
        Mbaracayú, Canindeyú, Paraguay.
46. *Linepithema pulex*, head, full face view.

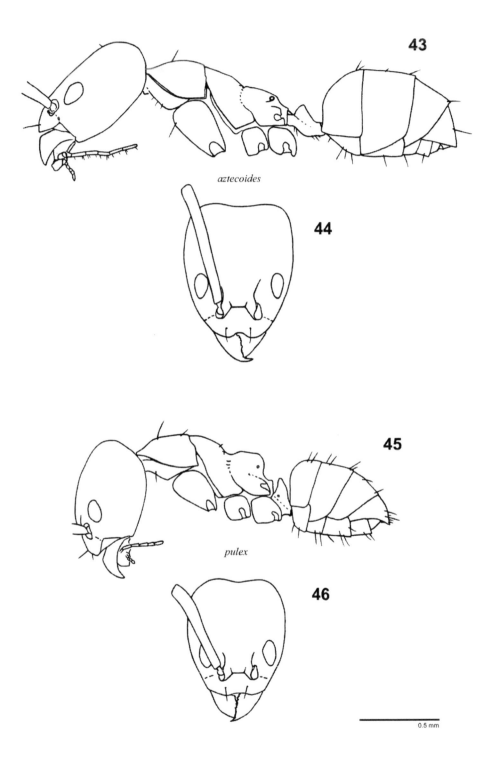

43

aztecoides

44

45

pulex

46

0.5 mm

Figures 47-54. Line drawings of *Linepithema* male specimens.
Abbreviations are as follows: *c*, cuspis; *dd*, distal arm of digitus; *dp*,
proximal arm of digitus; *g*, gonostylus; *Rs+M*, the confluence of
Radial sector and Median veins; *M.f2*, second free abscissa of the
Median vein; *smc1*, first submarginal cell; *smc2*, second submarginal
cell; *vp*, ventrodistal process of volsella.

47. *Linepithema tsachila*, fore wing, specimen from Maquipucuna,
       Pichincha, Ecuador.
48. *Linepithema humile*, fore wing, specimen from Victoria, Entre
       Rios, Argentina.
49. *Linepithema fuscum*, volsella, specimen from Guagua Sumaco,
       Napo, Ecuador.
50. *Linepithema humile*, volsella, specimen from Davis, California,
       USA.
51. *Linepithema tsachila*, body, lateral view, specimen from
       Maquipucuna, Pichincha, Ecuador.
52. *Linepithema micans*, body, lateral view, specimen from Tafi del
       Valle, Tucumán, Argentina.
53. *Linepithema dispertitum*, body, lateral view, specimen from
       Parque Nacional Armando Bermudez, Santiago, Dominican
       Republic.
54. *Linepithema humile*, body, lateral view, specimen from Victoria,
       Entre Rios, Argentina.

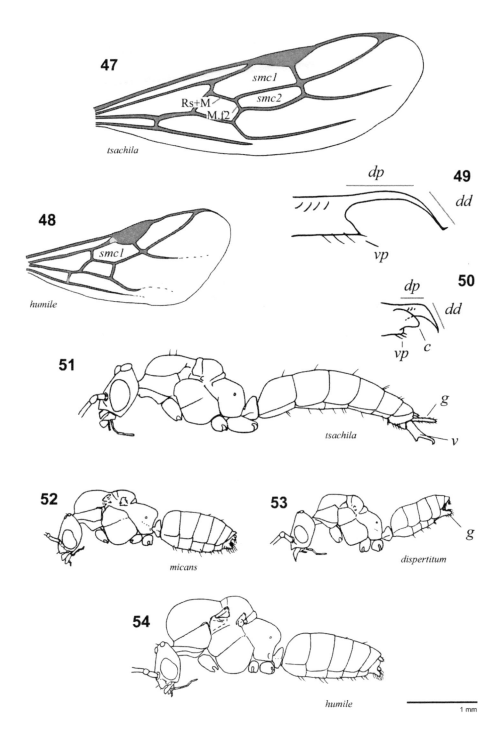

47 tsachila
smc1
smc2
Rs+M
M.f2

48 humile
smc1

49 dp dd vp

50 dp dd vp c

51 tsachila
g
v

52 micans

53 dispertitum
g

54 humile

1 mm

Figures 55-64. Line drawings of Fuscum-group male specimens.
Head and volsella figures of conspecifics are drawn from the same
specimens. Abbreviations are as follows: *dd*, distal arm of digitus;
dp, proximal arm of digitus; *vh*, height of volsella.

55. *Linepithema angulatum*, head, full face view, specimen from
        Misahualli, Napo, Ecuador.
56. *Linepithema angulatum*, volsella.
57. *Linepithema fuscum*, head, full face view, specimen from Guagua
        Sumaco, Napo, Ecuador.
58. *Linepithema fuscum*, volsella.
59. *Linepithema keiteli*, head, full face view, specimen from La Vega,
        Dominican Republic.
60. *Linepithema keiteli*, volsella.
61. *Linepithema piliferum*, head, full face view, specimen from
        Cosanga, Napo, Ecuador.
62. *Linepithema piliferum*, volsella.
63. *Linepithema tsachila* paratype, head, full face view. 64.
        Linepithema tsachila paratype, volsella.

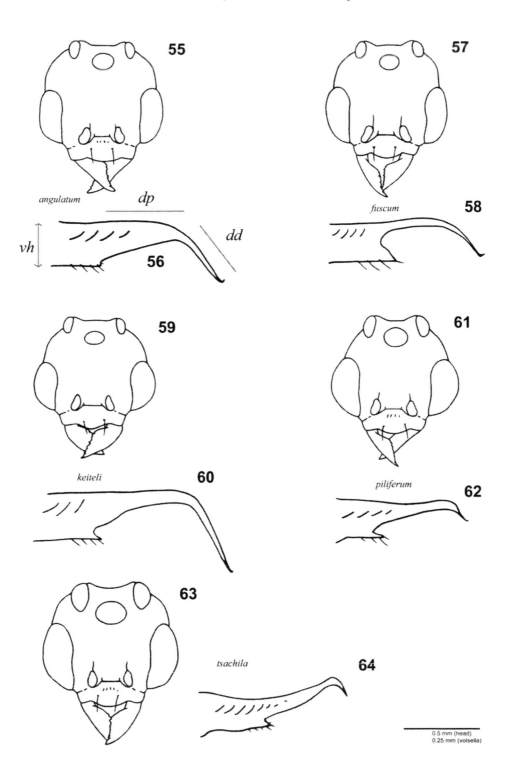

Figures 65-76. Line drawings of *Linepithema* male specimen heads in full face view.

65. *Linepithema neotropicum* paratype.
66. *Linepithema cerradense*, specimen from Buena Vista, Santa Cruz, Bolivia.
67. *Linepithema leucomelas*, specimen from Ilha dos Buzios, São Paulo, Brazil.
68. *Linepithema iniquum*, specimen from Baños, Tungurahua, Ecuador.
69. *Linepithema pulex* paratype.
70. *Linepithema dispertitum*, specimen from Parque Nacional Armando Bermudez, Santiago, Dominican Republic.
71. *Linepithema dispertitum*, specimen from Cuernavaca, Morelos, Mexico.
72. *Linepithema dispertitum*, specimen from 3 km SSE San Andrés Semetabaj, Sololá, Guatemala.
73. *Linepithema oblongum*, specimen from 7 km N Tafi de Valle, Tucumán, Argentina..
74. *Linepithema humile*, specimen from Victoria, Entre Rios, Argentina.
75. *Linepithema gallardoi*, specimen from La Falda, Córdoba, Argentina.
76. *Linepithema micans*, specimen from Campos do Jordão, São Paulo, Brazil.

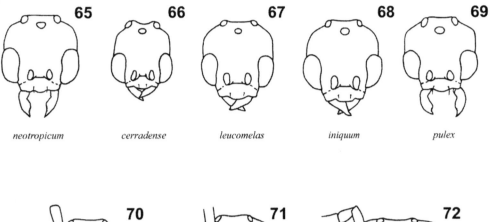

65 66 67 68 69

neotropicum *cerradense* *leucomelas* *iniquum* *pulex*

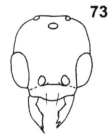
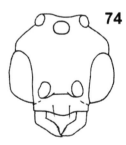
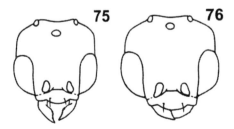

dispertitum (DR) *dispertitum (MX)* *dispertitum (GT)*

oblongum *humile* *gallardoi* *micans*

0.5 mm

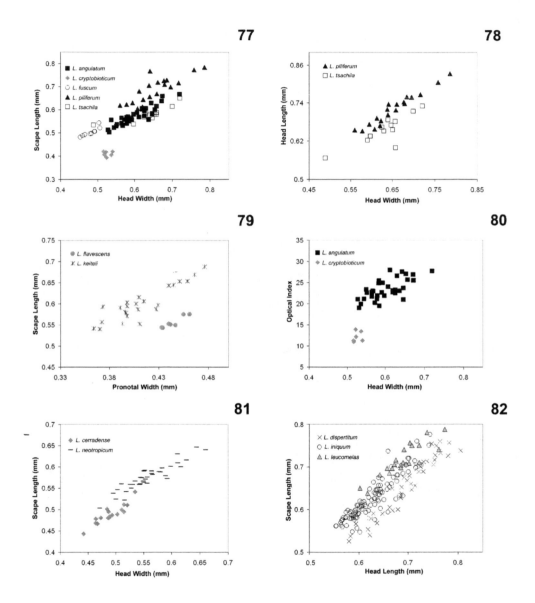

Figures 77-82. Bivariate morphometric plots of various characters in *Linepithema* worker specimens.

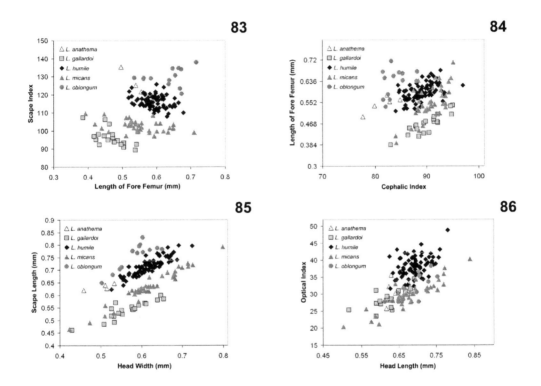

Figures 83-86. Bivariate morphometric plots of various characters in *Linepithema* worker specimens.

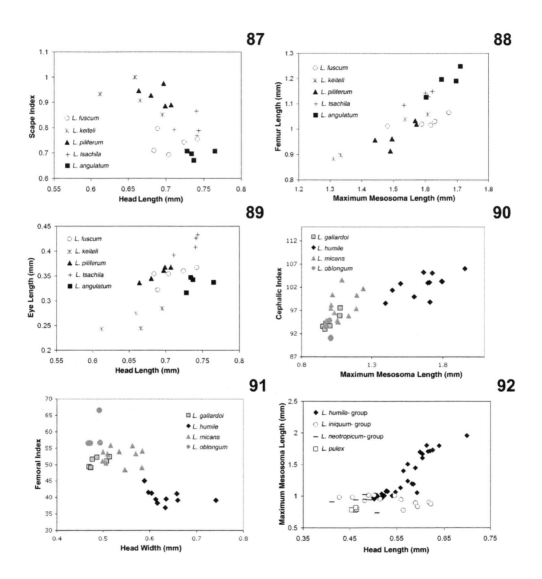

Figures 87-92. Bivariate morphometric plots of various characters in *Linepithema* male specimens.

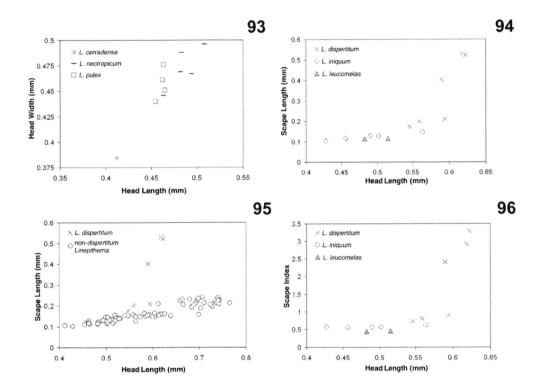

Figures 93-96. Bivariate morphometric plots of various characters in *Linepithema* male specimens.

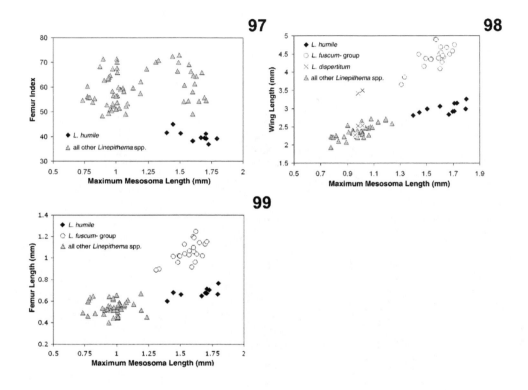

Figures 97-99. Bivariate morphometric plots of various characters in *Linepithema* male specimens.

Figures 100-103. Distribution maps of *Linepithema* species based on examined museum specimens.

100. *Linepithema angulatum, L. leucomelas, L. cryptobioticum.*
101. *Linepithema humile, L. oblongum.*
102. *Linepithema dispertitum, L. iniquum.*
103. *Linepithema tsachila, L. fuscum, L. micans.*

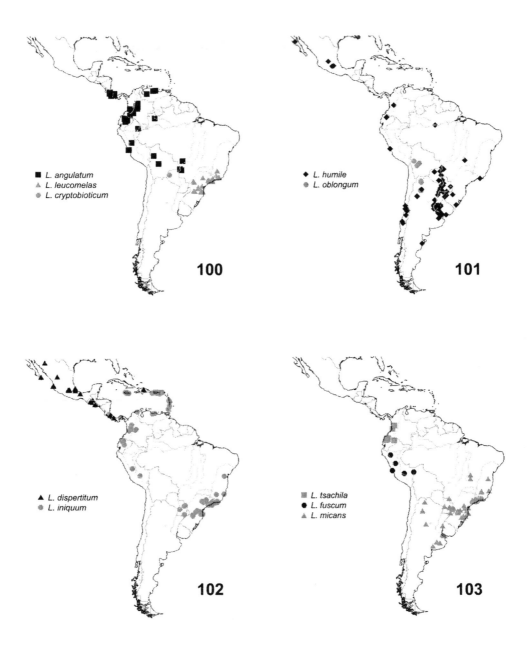

Figures 104-106. Distribution maps of *Linepithema* species based on examined museum specimens.

104. *Linepithema neotropicum, L. cerradense.*
105. *Linepithema piliferum, L. aztecoides, L. pulex.*
106. *Linepithema gallardoi, L. anathema.*

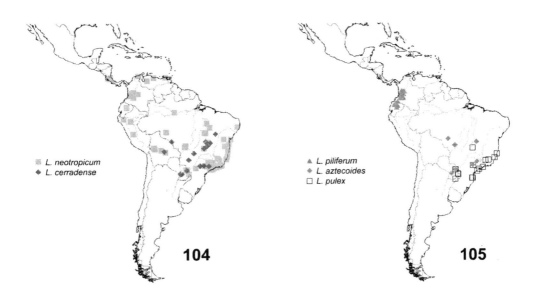

104

L. neotropicum
L. cerradense

105

L. piliferum
L. aztecoides
L. pulex

L. gallardoi
L. anathema

106

Figures 107-108. Distribution maps of *Linepithema* species based on examined museum specimens.

107. Distribution of *Linepithema* in the Caribbean region.
108. Worldwide distribution of all *Linepithema* species.

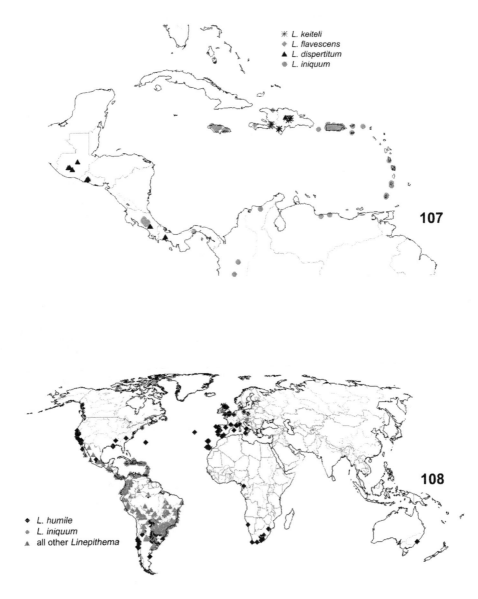

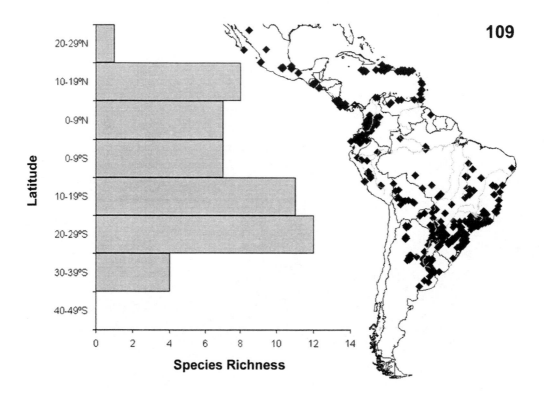

Figure 109. Species richness of *Linepithema* by 10 degree latitudinal increments over the native distribution, based on examined museum specimens. Introduced populations of *Linepithema humile* have been excluded.

Figure 110. *Linepithema aztecoides* workers from the Mbaracayú Reserve in Paraguay, showing the elevated gaster posture that is unique among *Linepithema* species.